T207 Engineering: mechanics, materials, design

The Open

BLOCK 3

DYNAMICS: GETTING MOVING

PARTS 3 AND 4

This publication forms part of an Open University course T207 *Engineering: mechanics, materials, design*. Details of this and other Open University courses can be obtained from the Student Registration and Enquiry Service, The Open University, PO Box 197, Milton Keynes MK7 6BJ, United Kingdom: tel. +44 (0)845 300 60 90, email general-enquiries@open.ac.uk

Alternatively, you may visit the Open University website at http://www.open.ac.uk where you can learn more about the wide range of courses and packs offered at all levels by The Open University.

To purchase a selection of Open University course materials visit http://www.ouw.co.uk, or contact Open University Worldwide, Walton Hall, Milton Keynes MK7 6AA, United Kingdom for a brochure: tel. +44 (0)1908 858793; fax +44 (0)1908 858787; email ouw-customer-services@open.ac.uk

The Open University
Walton Hall, Milton Keynes
MK7 6AA

First published 2004. Second edition 2006. Third edition 2009.

Edited and designed by The Open University.

Typeset in India by Alden Prepress Services, Chennai.

Printed and bound in the United Kingdom by Hobbs the Printers Limited, Brunel Road, Totton, Hampshire SO40 3WX.

ISBN 978 0 7492 2364 9

3.1

The paper used in this publication contains pulp sourced from forests independently certified to the Forest Stewardship Council (FSC) principles and criteria. Chain of custody certification allows the pulp from these forests to be tracked to the end use (see www.fsc-uk.org).

T207 presentation course team

Dr Jim Moffatt, course chair
Abi Lewis, course manager

T207 production course team

Professor Nicholas Braithwaite, course chair
Peta Jellis, course manager

Academic staff

Professor Adrian Demaid
Professor Chris Earl
Professor Lyndon Edwards
Mark Endean
Dr Michael Fitzpatrick
Dr Alec Goodyear
Dr Andrew Greasley
Dr Salih Gungor
Michael Hush
Jan Kowal
Dr Keith Martin
Dr Jim Moffatt
Dr Martin Rist
Dr Joe Rooney
Dr David Sharp
Graham Weaver
Dr George Weidmann

Support staff

Sylvan Bentley, picture researcher
Philippa Broadbent, materials procurement
Daphne Cross, materials procurement
Tony Duggan, learning projects manager
Vicky Eves, graphic artist
Garry Hammond, editor
Karen Lemmon, compositor
Vicki McCulloch, designer
Lara Mynors, media project manager
Lynn Short, software development

Part 3
Generating forces: electric motors

CONTENTS

1 INTRODUCTION

1.1 Aims

The overall aims of Part 3 of this block are:

- To provide an overview, from a historical perspective, of the ways of generating and transmitting forces, with special reference to electromagnetic forces.
- To identify the source of motive force in direct current (d.c.) electric motors.
- To describe d.c. permanent magnet motor characteristics in terms of torque, speed, power and efficiency.
- To identify basic d.c. motor design principles and relate these to more complex motors and electric machines.
- To introduce the fundamental principles of alternating current motors and generators.

1.2 Forces in electric motors

So far in your study of dynamics you have found out how to describe and quantify motion, forces and energy. Along the way you have met a number of devices, or machines, that are able to modify these quantities in order to do something useful. Examples have included toothbrushes, windscreen wipers and car engines. Broadly speaking, such machines can be thought of as 'energy converters' or *transducers*, changing energy from one form to another. That makes the human body a machine of sorts. From an engineering viewpoint, however, it is often more useful to think about forces. Machines that transmit or modify applied forces, like those listed above, can often appear quite complicated, but are really based on a small number of fundamental design principles. To emphasize this point, after a brief historical introduction to the diversity of force-modifying machines, this part of the Dynamics block will concentrate in detail on the generation and transmission of forces in the electric motor, from the simple small motor you might find in your electric toothbrush, to complex motors that are literally tens of millions of times more powerful.

2 THE SEARCH FOR POWER

Wouldn't it be wonderful if we could find a continuous source of mechanical power? Simple devices for using human muscle power more effectively have been around for a very long time. Archaeological evidence suggests that early humans were using basic tools around two million years ago. From the moment our ancestors set about modifying and improving their circumstances there was a need to hammer, cut and dig. To begin with, sharp flints were used for cutting and digging. Later we used animal hide to bind the flints to sticks, creating knives, axes and harpoons. In fact, sharp-edged toolheads of this type are actually one of the most basic type of machine – the *wedge*. The two inclined faces of a wedge convert the pushing force in one direction into two splitting forces which each have components at right angles to the blade.

Around the first century AD a Greek engineer and mathematician, Hero of Alexandria, developed a theory from the idea that all complex machines consisted of combinations of five simple machines:

the wedge

the lever

the pulley

the wheel and axle

the screw.

You'll be familiar with all of these from the earlier parts of Block 3; you might even want to add 'gears' to this list of simple machines – many people do. A more basic classification considers a wedge to be a double inclined plane and distinguishes only two classes of machine:

the lever

the inclined plane.

This is because the wheel and axle, and the pulley, both really work on the principle of a rotating lever, while the screw can be considered as an inclined plane wrapped around a core (Figure 3.1).

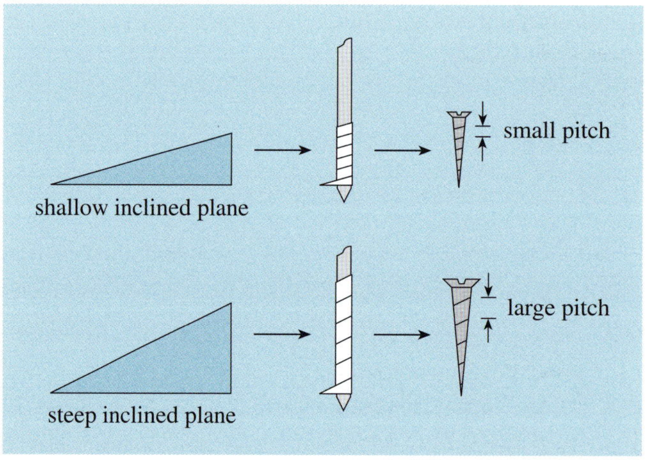

Figure 3.1 The screw thread as an inclined plane

You'll start to question this whole idea of basic machines when you try to envisage a modern electric motor (or even a human being) as a combination of levers, pulleys and inclined planes. Nevertheless, it is often a good idea to break down a relatively complicated mechanism into simpler components in order to understand how it works.

Exercise 3.1

A stone-age axe might seem 'simple' but it is actually a 'compound' machine. What are its basic components?

Exercise 3.2

Using the five machines on Hero's list, plus 'gears', identify the basic machines that make up the following more complicated machines:

(a) a hand-powered drill (Figure 3.2a)

(b) a two-handled corkscrew (Figure 3.2b)

(c) a car jack (Figure 3.2c).

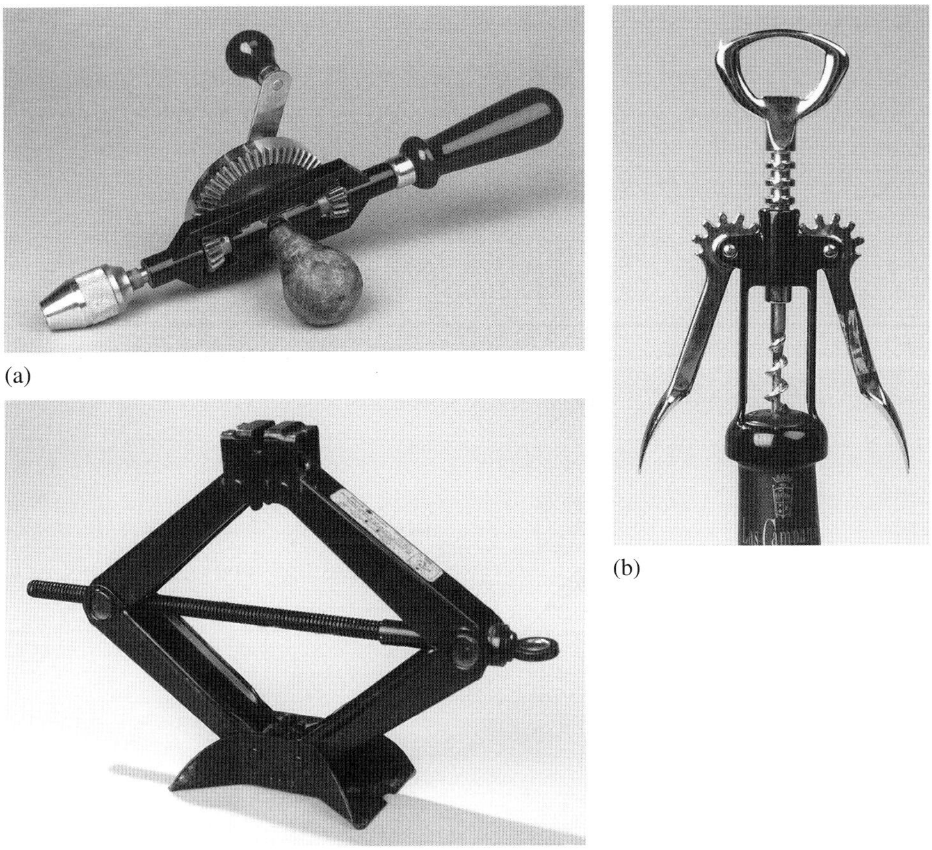

(a)

(b)

(c)

Figure 3.2 Examples of 'complex' machines: (a) hand-powered drill, (b) two-handled corkscrew, (c) car jack

As well as using axes, prehistoric artisans employed inclined planes to get underneath heavy loads, and also used levers to gain mechanical advantage when shifting large pieces of rock. Furthermore, by bending a flexible branch or a deer's antler and connecting the ends with sinew to make a bow, they were able to store energy for release in an arrow.

Somewhere, a few thousand years ago, humans hatched an idea – placing cylindrical rollers, probably fallen tree trunks, underneath large rocks could help them to move the rocks. Some think this principle was used in the construction of Stonehenge in southern Britain, and the Great Pyramids in

Egypt. The invention of the wheel and axle, most likely originating from the roller idea, is thought to have occurred some 5000 years ago.

About the same time, the first mechanism for harnessing natural power came into general use: the *sail*. This probably originated in one of the ancient river civilizations, possibly along the Nile, Euphrates or Yangtse, starting with a boatman simply raising a patch of cloth to the wind in order to hasten his journey. Some 2000 years ago the power of flowing water was also harnessed when the *water wheel* was invented in ancient Rome. A water wheel is the simplest form of *turbine*, a device in which a moving fluid drives a wheel or motor to convert the kinetic or potential energy of the fluid into mechanical energy. The *windmill* is a slightly more recent innovation, originating in the wind-blown salt deserts of Persia around the seventh century AD, when a paddle wheel was fixed to a vertical shaft and pushed into a millstone which could then rotate and grind cereal. By the fourteenth and fifteenth centuries water wheels and windmills were widespread across Europe, crushing ore, sawing wood, pumping bellows and drawing wire. Two centuries after this they were powering labour-saving devices that could knit, spin thread, weave cloth and produce blast for smelting furnaces.

Then along came a device that would instigate rapid and widespread industrialization, the *steam engine*. In fact, the idea of getting power from steam was not new – Hero of Alexandria, the deconstructor of complex machines, had already done it in ancient Greece. He described the construction of a rotating sphere driven by steam, called an 'aeolipile' (Figure 3.3) – but he treated it as a kind of curiosity and put it to no useful purpose.

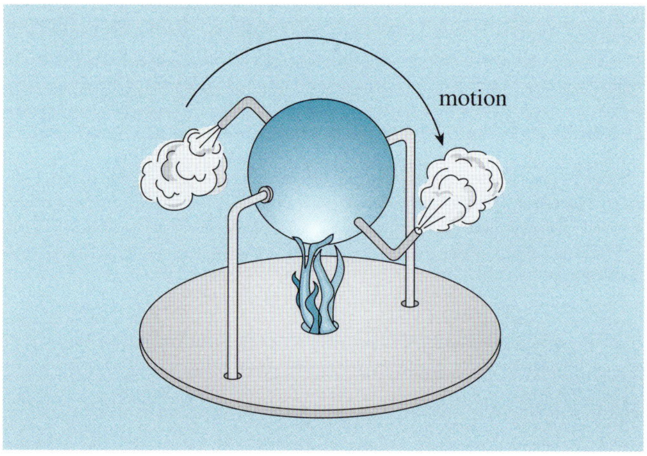

Figure 3.3 Hero's aeolipile – the earliest steam engine? Steam from boiling water inside the sphere is forced out through jets, causing the sphere to spin on its axis

Steam power was therefore ignored until around the beginning of the eighteenth century when primitive steam engines were used to pump water from coalmines. However, steam power didn't really take off until British engineer James Watt made a significant modification that vastly improved engine efficiency. In a steam engine, heat from a furnace is used to raise steam, the expansion of which forces a piston to move up and down in a cylinder to provide mechanical energy. In early engines, a jet of water was used to cool the cylinder after each stroke. Watt found that it was far more efficient to use a separate chamber, called a condenser, to cool the steam (Figure 3.4). His innovation allowed steam power to spark the British industrial revolution towards the end of the eighteenth century.

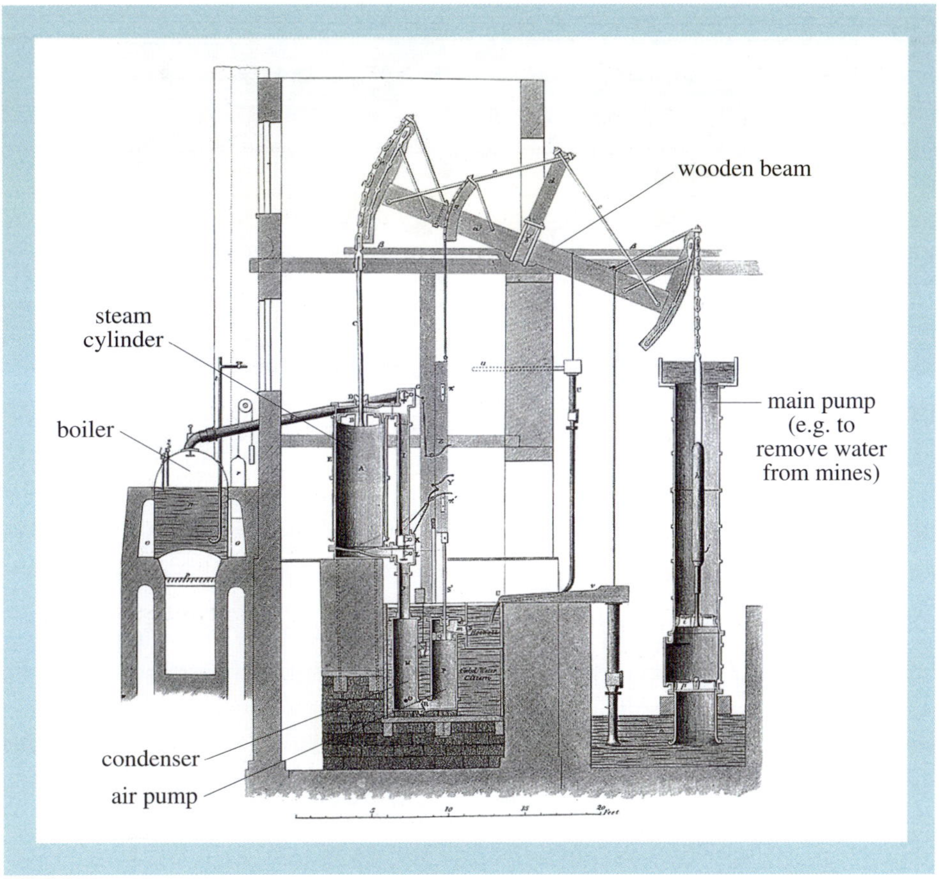

Figure 3.4 Watt's steam engine of 1774

In the first half of the nineteenth century, a means of converting electrical energy into mechanical energy appeared in the form of the electric motor – a device we'll examine in detail shortly. By the mid 1800s, electric motors were being used to drive large-scale industrial machinery and by the end of the century trains and trams, driven by electric traction motors, were to be found in many major cities and towns of Europe and North America. Small electricity-generating stations, initially using steam engines to drive alternating current generators or *alternators*, distributed electrical power locally to run these industrial motors. By the start of the twentieth century multi-stage steam-driven turbines replaced steam engines as the so-called 'prime-mover' in electrical power stations, and electricity began to be distributed on a larger scale. Fossil fuels were initially the primary energy source, but soon the energy of falling water was also harnessed in hydroelectric plants. (Block 4 will revisit this story.) Later, nuclear energy was also used. As electrical engineering developed, smaller motors were devised and by the 1920s motorized electrical appliances had invaded numerous homes, particularly in North America, in the form of vacuum cleaners and washing machines. Today, electric motors really are everywhere. I've counted at least a dozen in my kitchen alone, and many appliances contain a couple of motors. (I reckon there are *five* in a rather unattractive double-cassette/radio/CD-player sitting on my windowsill – two for tape transport, one each for CD rotation and laser movement, and another to open the lid!) Electric motors have become ubiquitous because they are relatively small, clean, quiet and efficient, and because they can be used to supply a huge range of power, from a fraction of a watt to several megawatts. So let's find out exactly how they work.

3 ELECTRICITY AND MAGNETISM: FORCES FROM NOWHERE?

Electric motors convert electrical energy into mechanical energy or, more specifically, into rotational kinetic energy in a revolving shaft. But how does 'electrical energy' make a shaft rotate? You might know that there is a magnet in there somewhere – perhaps that's got something to do with it? Let's start from the beginning.

3.1 Magnetic fields and how to picture them

Most people know how to use magnets to create motion. Just holding two bar magnets near to each other makes it clear that there is a force between them that has the potential to create movement. If you have used a bar magnet to pick up pins or paper clips you will know that they stick mainly to the two ends, or *poles*, of the magnet, where the attracting force seems to be greatest. You should also be aware that the poles are different because two bar magnets will repel or attract each other depending on which ends are brought together. These phenomena indicate that magnetism has both *magnitude* and *direction*. The magnetic poles can be distinguished by suspending a bar magnet, for example on a thread, so that it can swing freely in a horizontal plane. Eventually it will come to rest pointing roughly north–south: that's exactly how a compass works. By convention, the pole that points towards the north is called the 'north-seeking' or north (N) pole of the magnet, while the other is the south (S) pole. If you differentiate between poles in this way you will find that *like* poles repel (i.e. S and S, or N and N) while *unlike* poles attract (i.e. N and S).

For the time being we do not need to be too concerned about *why* materials are magnetic. Let's just use the idea that magnetic materials can 'feel' the presence of other magnetic materials. This is what we mean by 'magnetism'. The region around a magnet where it has a magnetic effect is called its *magnetic field,* and the direction and magnitude of this field can vary from place to place. For example, the overall curved symmetrical shape of the magnetic field around a bar magnet can be revealed using many small compasses, as shown in Figure 3.5(a).

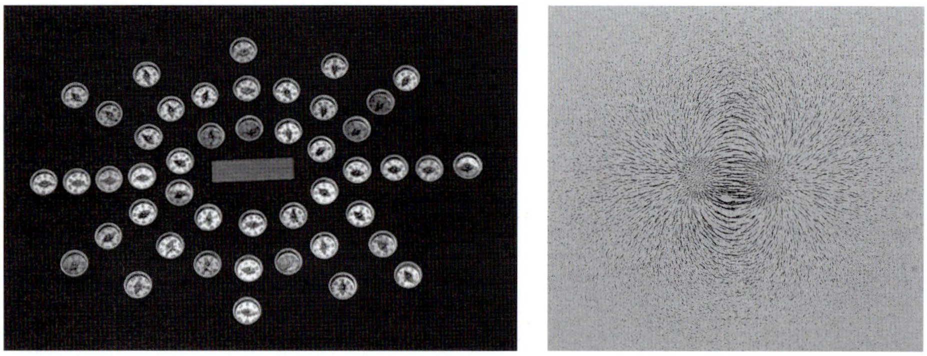

(a) (b)

Figure 3.5 (a) the direction of the magnetic field around a bar magnet revealed using many small compasses, (b) iron filings sprinkled on a piece of paper placed over a bar magnet

The field pattern emerges because, as I've already indicated, the needle of a compass is just a small, weak magnet that lines itself up with the local direction of the prevailing magnetic field, wherever it is placed. A similar effect can be achieved using iron filings sprinkled on a piece of paper placed over the magnet (Figure 3.5b). In this case the magnetic field *magnetizes* the iron filings so that they turn into tiny magnets; each individual filing acts like a miniature compass needle.

But what is actually going on here? How do the forces that arise from a magnetic field bring a compass needle to point in the 'right' direction? To answer these questions it is a good idea to adopt a means of picturing the *effect* of a magnetic field. Michael Faraday, a brilliant nineteenth-century experimentalist and one of the great discoverers in magnetism and electricity, devised a way to do this.

Faraday came up with the idea of drawing *magnetic field lines*, which he called 'lines of force', to show how an external magnetic field might influence another magnet. By convention, these field lines are drawn such that they show the direction in which an 'imaginary' *north* magnetic pole would move if it were placed in the field. Let's start by considering a *uniform* magnetic field, i.e. one whose direction and strength is the same everywhere. Such a field is depicted in Figure 3.6(a) using a set of straight parallel lines with direction arrows that all point towards the top of the page.

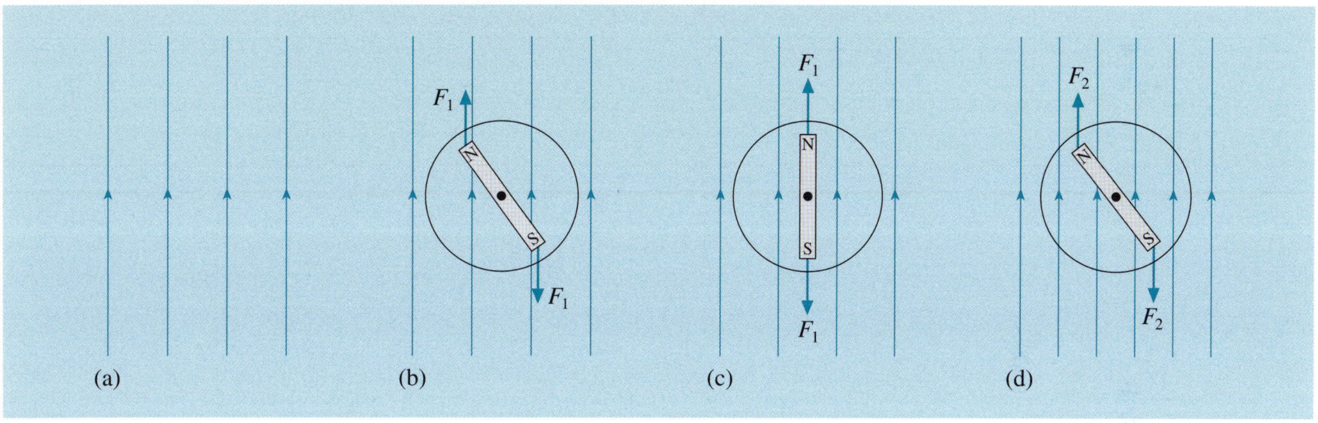

Figure 3.6 Torque on a compass needle in a uniform magnetic field

A north magnetic pole placed in this field would experience a force in the direction of the arrows – upwards. Conversely, a south magnetic pole would feel a force in the opposite direction – downwards. Now look at Figure 3.6(b) which shows a compass needle, actually a tiny bar magnet, oriented at an angle to the uniform magnetic field. The north pole of the needle wants to move towards the top of the page, in the direction of the field line arrows, while the south pole of the needle feels a force that makes it move in the opposite direction. In other words, the needle experiences a *torque* about its point of rotation. The torque is zero when the needle is aligned with the uniform external magnetic field, as shown in Figure 3.6(c), at which point it remains at rest.

Magnetic field lines are often called *lines of magnetic flux*. Although we now consider field lines to be an abstract means of describing the magnetic effect, Faraday and other scientists of his time thought that they represented something more tangible. They thought that the physical properties of space might be altered by the presence of the field in some way, or that something was flowing, hence the term 'flux', which has persisted to this day.

The magnetic field in Figures 3.6 (a) – (c) is such that it generates a force F_1 on each pole of the compass needle magnet. But suppose a different magnetic field was used that generated a greater force, say F_2. This is simply represented by drawing the lines of flux closer together, as shown in Figure 3.6(d). In other words it is the *flux density* – the number of lines per unit area – that represents the size of the magnetic force. Hence field lines give information about both the *direction* of magnetism and its *relative magnitude*.

Next, we will consider a *non-uniform* field, one whose direction or magnitude varies from place to place. In particular, if the direction of force varies then the lines of magnetic flux must curve. The non-uniform field patterns depicted in Figure 3.5 have been redrawn using conventional magnetic field lines in Figure 3.7. Note again that the field lines are marked with a direction which is the same as that in which the north pole of a compass needle would point: towards the south pole and away from the north pole.

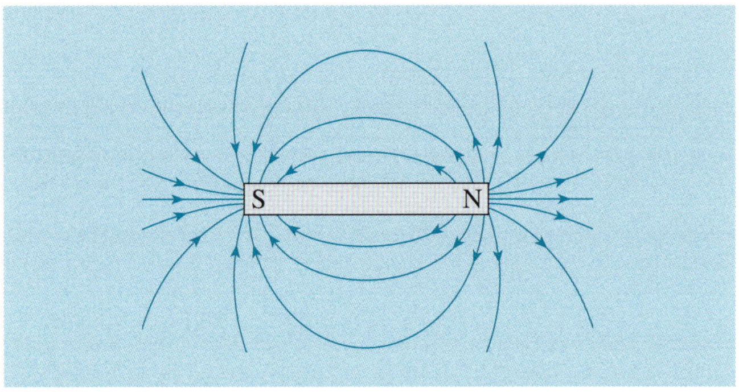

Figure 3.7 Overall field of a bar magnet depicted by drawing field lines

Let me just make that last point quite clear: by convention the 'north-seeking' pole of a compass needle points to the *south* magnetic pole! This doesn't contradict your common knowledge of a compass pointing 'north' because the Earth actually acts as if it contains a giant upside-down bar magnet with its south *magnetic* pole close to what we call *geographic* north; see **▼The Earth's magnetic field▲**.

▼The Earth's magnetic field▲

William Gilbert, who once worked as a physician to Queen Elizabeth I, discovered the nature of the Earth's magnetic field over 400 years ago. For many years Gilbert had experimented with a naturally occurring magnetic material called lodestone, which is a magnetic iron oxide (Fe_3O_4), and also with artificially magnetized iron. He constructed a spherical magnet, which he called a terrella or 'little Earth', and investigated its properties by moving a small compass needle around it. He found that the needle always pointed towards a pole when it was placed in a plane tangential to the sphere surface (as you would hold a compass when standing on the Earth). He also found that it pointed downwards at an angle to the sphere surface when turned on its side – see Figure 3.8(a). He concluded that the Earth itself was a giant magnet. This isn't strictly true, but it does have a magnetic field similar to that of a bar magnet (Figure 3.8b). Note that the south magnetic pole appears to be located near the geographical North Pole.

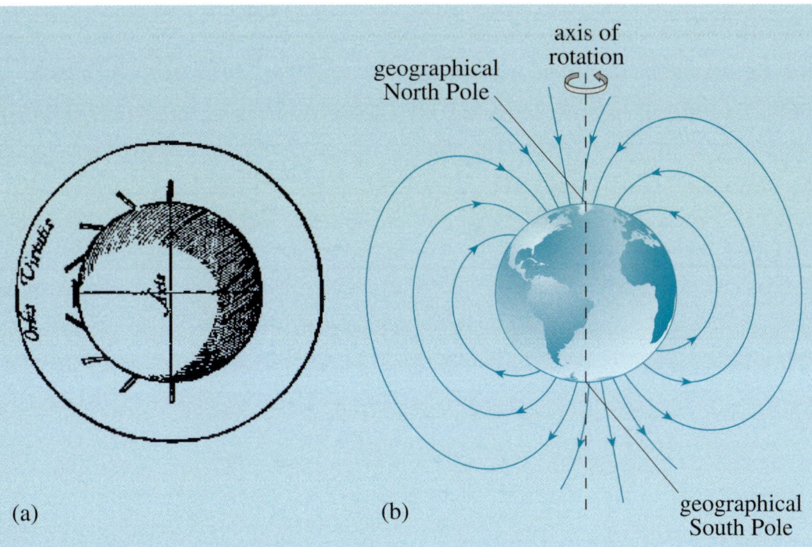

Figure 3.8 (a) Terrella, a model of the magnetic Earth from *De Magnete* by William Gilbert (published 1600), (b) the Earth's magnetic field

The Earth's magnetic field doesn't arise from fixed magnetic materials but is thought to be associated with electrical currents produced within the moving mass of molten iron and nickel that makes up the Earth's outer core (surrounding a solid inner core), between about 2800 km and 5000 km below the surface. The field is also influenced to some extent by charged particles arriving from the Sun. Because the field isn't derived from a solid permanent magnet, the exact location of the north magnetic pole varies. It is currently over 1500 km from the geographic north pole and drifting at about 40 km per year. Furthermore, it wobbles about its mean position, following an elliptical course that shifts it by up to 80 km a day. The Earth's magnetic field has even undergone complete reversals, during which the north and south poles switch round. This has happened over 60 times during the last 20 million years, the most recent being about 250 000 years ago.

On a global scale, it is clear from Figure 3.8(b) that the Earth's magnetic field lines are curved and that their spacing varies i.e. the Earth's field is non-uniform. But on a much smaller scale the field lines appear straight and the variation in the strength of the field is minimal. In fact when using a compass, or when doing laboratory-scale experiments, the Earth's magnetic field can be considered to be entirely uniform. The field lines drawn in Figure 3.6, showing how a compass works, could be those of the Earth's magnetic field.

One important thing you should notice from looking at Figure 3.7 is that the magnetic field lines don't cross. In fact they *never* cross, because it would imply that the magnetic field has two directions at the point of intersection. This effectively means that field lines 'bunch up' near the poles of a magnet and are more spread out elsewhere. In other words the magnitude of the magnetic field (as measured by the flux density) is greater near the magnetic poles, and has a greater magnetic force associated with it.

Exercise 3.3

Look again at the field of a bar magnet, redrawn in Figure 3.9, and consider points **a** to **g** marked on the figure.

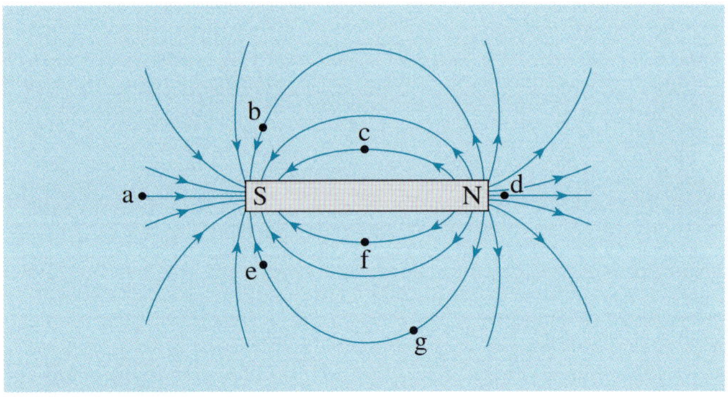

Figure 3.9 Magnetic field of a bar magnet

(a) At which of the points is the magnitude of the magnetic field the greatest?

(b) At which of the points is the magnitude of the magnetic field the smallest?

(c) For which pair of points are the magnitude and direction of the magnetic field identical?

(d) For which pair of points does the magnetic field have the same magnitude but different direction?

(e) For which pair of points does the magnetic field have the same direction but different magnitude?

Now, from what you have learnt so far, you might have noticed that a magnetic field doesn't generate a force on its own – it is actually the interaction of different magnetic fields that is important. In fact, when two magnets are brought together, the total, or *resultant*, magnetic field is a combination of the individual fields. This is why compass needles are deliberately weak magnets so they don't disrupt the field they are supposed to be measuring. You can see the different resultant fields that arise between two bar magnets when they are arranged to attract and repel, in Figures 3.10(a) and 3.10(b) respectively.

If the faces of two unlike magnetic poles are brought close enough together, or are sufficiently large and strongly magnetic, then a fairly uniform field exists between them, i.e. a field in which the lines of flux are equally spaced and point in the same direction – see Figure 3.10(c). You will meet this type of field when examining the ideas on which electric motors are based. Finally, be aware that all the fields that I have drawn so far are only two-dimensional representations of three-dimensional fields (Figure 3.10d).

In principle it would be useful to create a force, and hence movement, from the interacting fields of two magnets. The problem is that the movement stops when the magnets get too far apart, or if they collide with each other. One solution might be to keep moving the magnets about, perhaps even rotate them so that repulsion continuously turns into attraction. However, this requires a lot of effort, particularly when attempted on a large scale. Fortunately there's another way to create a magnetic field besides using permanent magnetic materials.

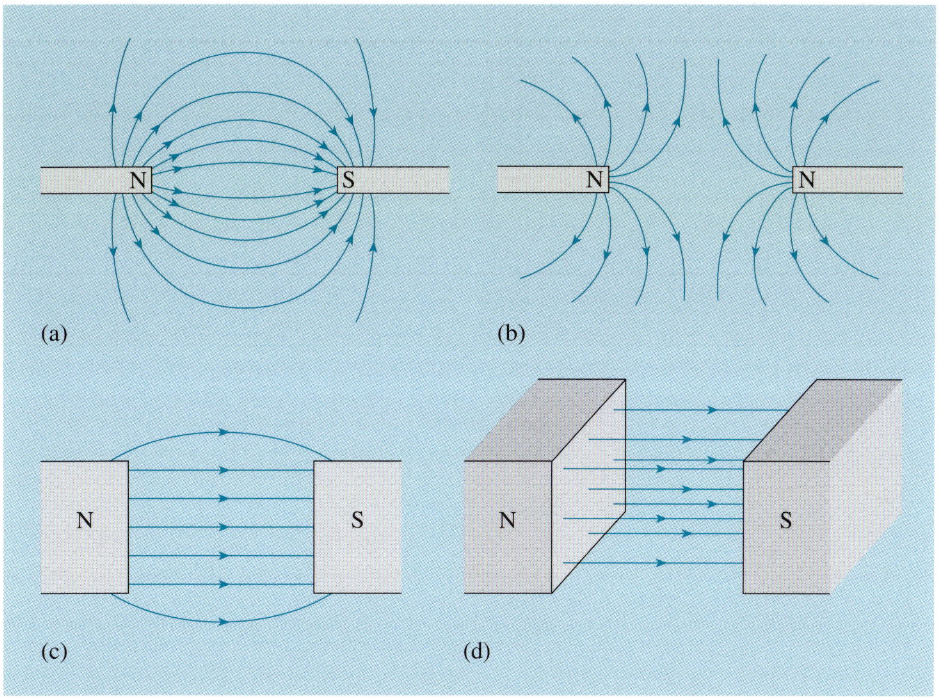

Figure 3.10 (a) field between bar magnets arranged to attract, (b) field between bar magnets arranged to repel, (c) uniform field in 2D, (d) uniform field in 3D

3.2 Electromagnetism: an accident waiting to happen

It is one of the fundamental principles of modern physics that electricity and magnetism are inextricably linked: every electric current also has a magnetic field. This branch of science, *electromagnetism*, provides the basis for understanding devices that allow engineers to generate forces and motion from electrical energy. Electromagnetism was probably first discovered in 1802 by Gian Domenico Romagnosi, an Italian philosopher and part-time scientist, who noticed that a magnetic needle could be made to move by placing a wire, connected to a battery, near to it. However, Romagnosi didn't understand the importance of his observation. He merely mentioned it in an obscure Italian publication, and his finding was ignored. It was some time before electromagnetism was 'rediscovered', again accidentally, by Danish physicist Hans Christian Oersted while preparing a student lecture in 1819. Oersted was getting ready to demonstrate the heating of a platinum wire connected to a voltaic cell (an early battery) when he noticed that the needle of a compass next to it spun round whenever the current was turned on and off. The compass needle, mounted on a wooden stand (Figure 3.11a), had only been intended for some other general demonstrations of magnetism. Although Oersted couldn't explain fully what he had observed he nevertheless knew it was significant and made sure he published his findings in important academic journals, ensuring he would get the credit later on – a lesson for us all! In fact, we now know that by moving a compass around a wire carrying a strong electric current it can be shown that the associated magnetic field is circular (Figure 3.11b).

Oersted also demonstrated that a current-carrying wire, when free to move, could be deflected by a magnet: see Figure 3.12. In other words a *force* is created that leads to *movement* of the wire. Again, this occurs because of two interacting magnetic fields: one from the wire, the other from the magnet.

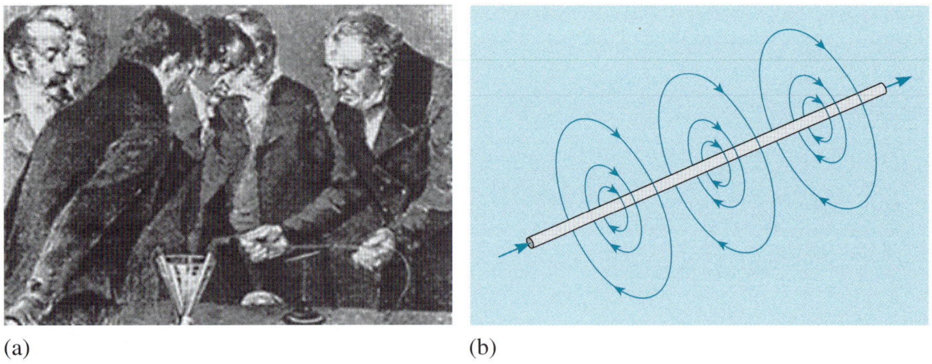

Figure 3.11 (a) Oersted experimenting with a compass and a current-carrying wire, (b) current-carrying wire and associated circular magnetic field

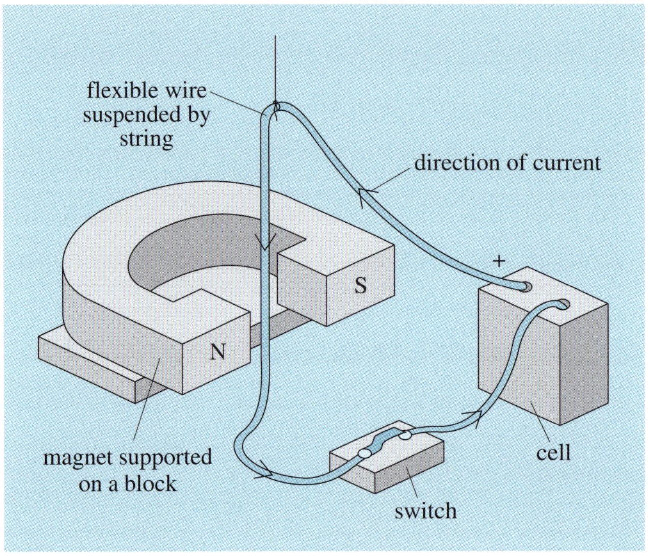

Figure 3.12 Simple electromagnetic experiment, in which the wire moves when the current is switched on

By now there was a flurry of activity in the scientific community around the link between electricity and magnetism. Shortly after Oersted's discoveries Peter Barlow, an English mathematician and engineer, showed how a metal disc with a current passing through it could be made to rotate in a magnetic field. His device, known as Barlow's wheel, was possibly the world's first electric motor. These important scientific discoveries are fundamental to our understanding of modern electric machines, so we will examine them in a little more detail.

3.3 An electromagnetic catapult

Let's look again at the field associated with a current-carrying wire. In Figure 3.13(a) I have drawn a wire in which the direction of conventional electric current (i.e. the flow from 'positive' to 'negative') is directly *into* the page. The magnetic field lines show the direction a compass would point if it were placed at various positions around the wire; note the field lines point *clockwise* around the wire.

If the electric current was reversed, so that it flowed *out of* the page, the field lines would also be reversed and point *anticlockwise*. You can remember the way this works by imagining that you are turning a corkscrew along the direction of conventional electric current: the corkscrew rotates in the same direction as the magnetic field lines point. This is called the *corkscrew rule*.

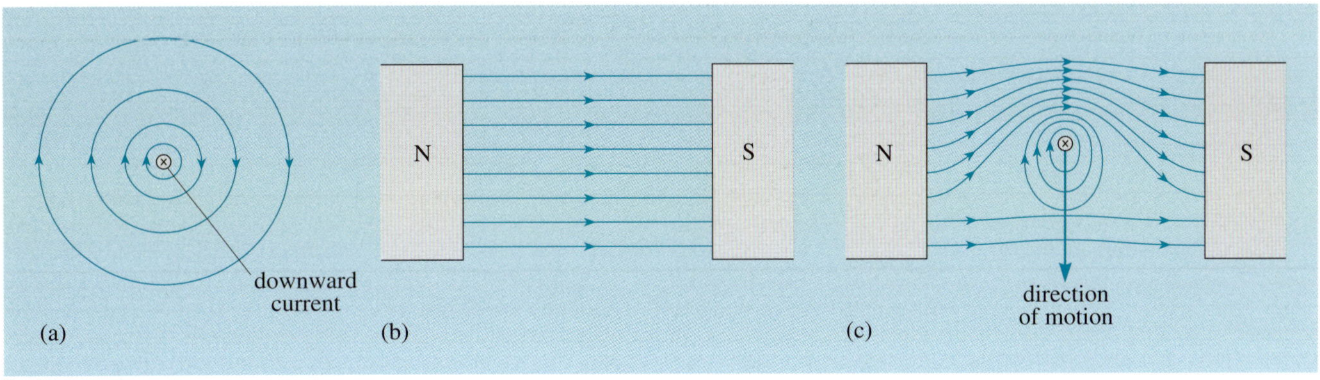

Figure 3.13 The motor effect: (a) field around current-carrying wire into page, (b) uniform magnetic field, (c) combined fields

The sketch of the magnetic field between flat north and south magnetic poles in Figure 3.13(b) shows field lines pointing straight from the north pole to the south pole. So what would happen if we were to combine the magnetic field of the wire with that of the magnet? Imagine that the wire is introduced into the uniform field and the current switched on. You should realize that there are regions where the two sets of field lines point in roughly the same direction (above the wire) and regions where they point in opposite directions (below the wire). As a result, the magnetic fields tend to reinforce each other above the wire and partly cancel each other out below it. The combined, or resultant, magnetic field is shown in Figure 3.13(c). Recall that closer-spaced field lines indicate a stronger field, so the figure indicates that the field is strengthened above the wire and weakened below it. The wire experiences a force tending to push it out of regions of high flux density and towards regions of low flux density. Hence the wire in Figure 3.13(c) must accelerate downwards. We call this the *motor effect* or, more descriptively, the *catapult effect.* Basically, electric motors work because different magnetic fields can be made to interact with each other in this way to create motion.

Exercise 3.4

(a) Look back at Figure 3.12, showing a wire hanging within the field of a horseshoe magnet. In which direction is the wire propelled when the current is switched on?

(b) Still considering the experiment shown in Figure 3.12, how do you think it might be possible to increase the force created?

If you had trouble with the first part of Exercise 3.4, take a look at Figure 3.14 and compare it with Figure 3.12.

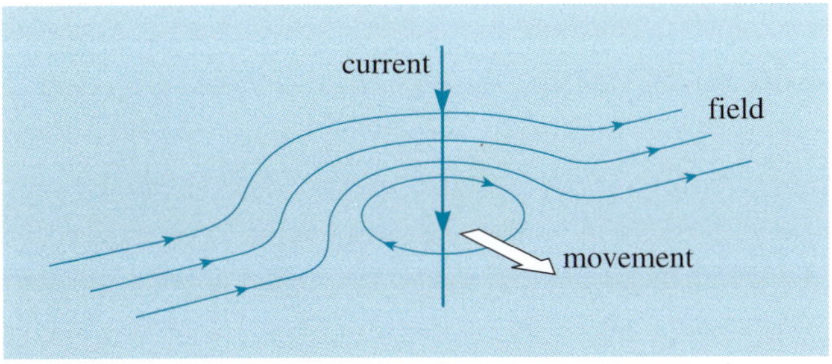

Figure 3.14 Sketch of fields to help understand motion in Figure 3.12

3.4 Quantifying electromagnetic force

When people first started to play around with apparatus like that shown in Figure 3.12 they soon discovered that they could create a greater force if they used a stronger magnet or if they increased the flow of electric current. Furthermore, simply using a larger magnet, i.e. one whose magnetism merely extended over a larger area, must also create more force because the field will act on a longer length of wire. We can say that there are four variables in the experiment: the magnetic field, the electric current, the length of wire and the force. Now you should know how to quantify the last three of these variables: electric current is measured in *amperes*, length in *metres* and force in *newtons*. In fact, if we were to measure these quantities we would find that increasing or decreasing either the electric current (I) or the wire length (l) by a certain factor, while using the same permanent magnet, would cause the force (F) to change by exactly the same factor. In mathematical terms we can write:

$$F \propto I\,l$$

where the symbol '$\propto$' means 'is proportional to'. But what about the magnetic field? Well, scientists have decided to describe this 'unknown' using a vector quantity called *magnetic flux density*, labelled $\boldsymbol{B}$. Here, we are just concerned with the magnitude of this vector, B, which is *defined* in terms of the three 'known' variables F, I and l:

$$B = \frac{F}{I\,l} \qquad (3.1)$$

Or, in words, magnetic flux density is the 'force per unit length per unit current' on a current-carrying conductor at right angles to a magnetic field. If we rearrange Equation (3.1) we get:

In these equations B represents the external applied field.

$$F = BIl$$

so the definition for B is chosen to be consistent with the known proportionality between F, I and l.

You have already seen, in Section 3.1 of this part of the block, how magnetism has both magnitude and direction, and that one way of representing it is to use magnetic field lines or *lines of magnetic flux*. (Remember, B is actually the magnitude of a vector, $\boldsymbol{B}$, whose direction is taken to be parallel to the direction of the magnetic field lines of the external applied field.) The use of the term *density* is consistent with the convention for drawing lines of flux such that their density is proportional to B, i.e. of drawing lines of flux closer together where the magnetic effect is strong, and further apart where it is weak.

Exercise 3.5

From Equation (3.1), write down appropriate units for B.

In fact, magnetic flux density is such an important quantity that it has its own SI unit, the *tesla* (T) – $1\ \mathrm{T} = 1\ \mathrm{N\ A^{-1}\ m^{-1}}$. The unit is named after Nikola Tesla, a Serbian electrical engineer and inventor whose most significant achievement was to develop ways of transforming electricity to and from high voltages, allowing it to be transmitted efficiently.

The tesla is a large unit. Small permanent magnets have a magnetic flux density of around 0.2 T, while large laboratory electromagnets, which use electric currents to create a field, can be of the order of 10 T. The value of B associated with the Earth's magnetic field is only 0.00005 T.

We just need to take this analysis a little further before we move on because there is one very important detail that I haven't yet pointed out. So far, I have assumed that the current flows in a wire that is at right angles to a uniform external magnetic field, as illustrated in Figure 3.12. However, if you were to take this wire and rotate it at right angles to the field lines, as shown in Figure 3.15, you would find that the force becomes weaker until, when the wire is parallel to the external field, the force is zero.

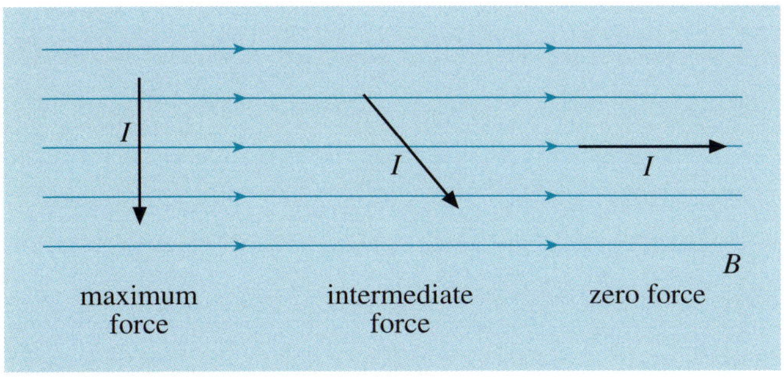

Figure 3.15 Rotating a conductor in a magnetic field

This occurs because magnetic flux density is a vector. It is the component of *B perpendicular* to the wire that creates the force. Hence, if the conductor cuts across the field lines at an angle θ, as shown in Figure 3.16(a), we need to consider the field component $B \sin \theta$ at right angles to the wire. The force on the wire is then given by:

$$F = (B \sin\theta) Il = BIl \sin\theta \qquad (3.2)$$

which is an important equation that later on will help you understand how a motor works. Note that the direction of the force is at right angles to both the field and the current as shown in Figure 3.16(b).

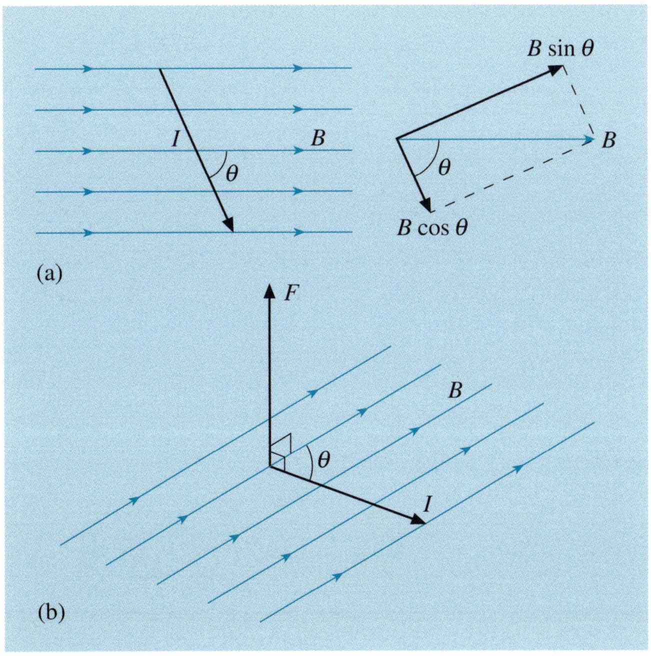

Figure 3.16 (a) resolving B at right-angles to the current I, (b) force is at right angles to both B and I

Example

In an experiment similar to that shown in Figure 3.12 a wire is placed at right angles to the lines of magnetic flux between the poles of a horseshoe magnet. If the length of wire in the field is 5.0 cm then it experiences a force of 0.02 N when a current of 2.0 A is passed through it. What is the magnitude of the magnetic flux density?

Solution

This looks like a job for Equation (3.2):

$$F = BIl\sin\theta$$

We know that:

$$F = 0.02 \text{ N}, \quad I = 2.0 \text{ A}, \quad l = 0.05 \text{ m}, \quad \theta = 90° \quad \text{so} \quad \sin\theta = 1$$

What we don't know – the thing we are looking for – is the magnetic flux density, B. So, making simple substitutions:

$$B = \frac{F}{Il\sin\theta} = \frac{0.02}{2.0 \times 0.05 \times 1} = 0.2 \text{ T}$$

SAQ 3.1 (Outcome 3.1)

As part of the experiment described in the above example the wire is rotated so that it is at 45° to the lines of flux. What is the new force on the wire?

SAQ 3.2 (Outcome 3.1)

In another experiment using a larger magnet, a 10-cm wire is placed vertically in a horizontal uniform magnetic field of magnitude 0.3 T.

(a) Calculate the force on the conductor when a current of 4.0 A is passed through it.

(b) Through what angle must the wire be turned in a vertical plane so that the force on it is halved?

Now let us take a moment to assess how much of an improvement the creation of an electromagnetic force is over a purely magnetic one, at least with respect to our ultimate goal of generating continuous mechanical power. Of course, if we can adjust the current then we have much more control over the force we create. But if the wire is catapulted out of the field how do we get it to come back again? Turning off the current to the wire immediately removes its associated magnetic field, but does this help?

Exercise 3.6

Think about what would happen to the wire in Figure 3.12 if we switched off the current while the wire was moving and still within the field of the permanent magnetic poles. What would happen and why? What would happen if we suddenly reversed the direction of the current? (Assume the wire stays vertical).

Used perhaps more as a lever than a catapult, this electromagnetic force aligns the read–write heads in disk drives during **▼Fast track changes▲**.

▼Fast track changes▲

Read–write heads in hard disk drives must be able to reach from the outermost radius to the innermost radius of the data area on the disk in about 7 ms.

Inside the disk drive, the head scans across the disk, driven by an electromagnetic force, and the original designers must have been faced with a challenge along the lines of 'how big a coil, carrying how much current is required for this task?' Let's see if we can find some solutions.

Figure 3.17 shows a schematic of the arrangement. The plan is to provide a constant torque for an interval of time to accelerate the head to halfway across the disk; this is immediately followed by an equal interval of reverse torque during which the head is brought to rest. You've already seen this kind of speed profile used for lifts and satellite orientation.

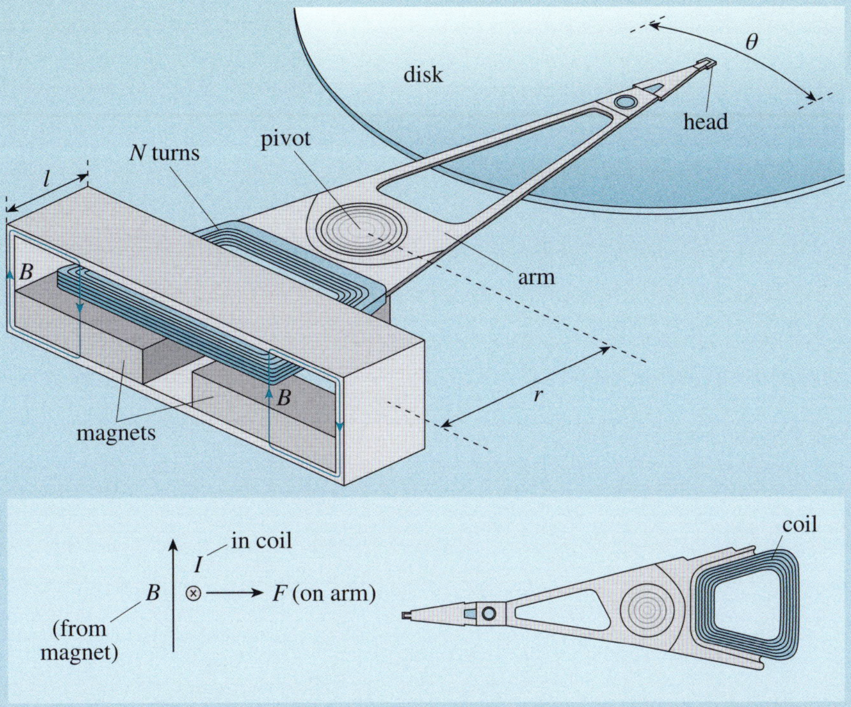

Figure 3.17 Schematic of the major components of the hard disk drive

There are three steps to follow, one from each of the parts in this block so far. From Part 1, we can describe the position of the head after being accelerated from rest at a constant rate, α rad s^{-2}, for t seconds:

$$\theta = \frac{1}{2}\alpha t^2$$

I_M is a moment of inertia, not a current!

We can rearrange this equation to find the acceleration needed to obtain a given angular displacement in a given time:

$$\alpha = \frac{2\theta}{t^2}$$

From Part 2 we have the rotational equivalent of $F = ma$ which is:

$$\Gamma = I_M \alpha$$

We can combine this with the angular displacement expression; to obtain a given constant angular displacement in a given time requires a constant torque:

$$\Gamma = \frac{2 I_M \theta}{t^2} \quad \text{(what is wanted)}$$

From earlier in this part of Block 3 we can express the torque arising from a force on N conductors in a coil (two sides) carrying current I, on average a distance r from the pivot point in a perpendicular magnetic field of flux density B, as:

$$\Gamma = 2NBIlr \text{ (what you can have for a given current and field)}$$

Equating what you want to what you can have:

$$2NBIlr = \frac{2 I_M \theta}{t^2}$$

relates the current we need, I, to seven other factors, all of which are specified, known or easily estimated (see Figure 3.17).

Angular displacement θ / rad	0.25
Acceleration time t / s	3.50×10^{-3}
Moment of inertia I_M / kg m^2	4.00×10^{-8}
Number of turns in coil N	100
Magnetic flux density B / T	0.200
Active coil length l / m	0.010
Effective radius of action r / m	0.015

$$I = \frac{I_M \theta}{NBlrt^2}$$
$$= \frac{4 \times 10^{-8} \times 0.25}{100 \times 0.2 \times 0.010 \times 0.015 \times (3.5 \times 10^{-3})^2}$$
$$= 0.27 \text{ A}$$

That's a reasonable amount of current to have to supply first one way, then the other, for 3.5 ms each time in order to move the head from rest on the outermost radius to rest on the innermost radius. You can probably see too that half the current in twice as many turns would be as good. Alternatively, if only a half-strength magnet is available then we'll need a longer coil or more turns to compensate, and thus we have solved the designers' question.

3.5 Summary of Section 3

- A magnetic field is the region around a magnet where it has a magnetic effect.
- Magnetic fields can be pictured using lines of magnetic flux that indicate both the direction and relative magnitude of the field.
- An electric current has a magnetic field associated with it.
- The interaction of the circular magnetic field of a current-carrying wire and the uniform magnetic field of a permanent magnet can give rise to a force; this is the 'motor effect'.
- Magnetic fields can be quantified using a vector called the magnetic flux density, $\boldsymbol{B}$.
- The electromagnetic force, F, that arises because of the motor effect depends on the magnitude of the magnetic flux density, B, the current, I, the length of wire, l, and the angle between the magnetic field and direction of current flow in the wire, θ, such that $F = BIl \sin \theta$.

4 A SIMPLE D.C. ELECTRIC MOTOR: PRINCIPLES AND DESIGN

4.1 Barlow's wheel

Web pages on the T207 CD-ROM describe how to construct a simple d.c. motor.

In 1822 Peter Barlow built a device widely considered to be one of the first prototypes of an electric motor supplied by a continuous current. Barlow's wheel is a kind of d.c. motor because it converts electrical energy into movement or kinetic energy. In fact, it's a matter of some debate as to who 'invented' the first electric motor. Many people consider the famous experimentalist Michael Faraday to have done so when he demonstrated how a current-carrying wire could be made to revolve around a fixed magnet as early as 1821.

Barlow's original model consisted of a small, star-shaped, brass wheel on an axle held in such a way that its lower extremities just dipped into a wooden trough of mercury below it (see Figure 3.18a). A horseshoe-shaped magnet was placed just above the bath of mercury so that the lower part of the wheel was in its magnetic field. Leads from a battery were then attached to the centre of the wheel (via its axle) and to the mercury bath, making a circuit in which current would flow mainly vertically along the wheel radius. Barlow found that the wheel often needed a little push to get it started, but that it would then rotate continuously.

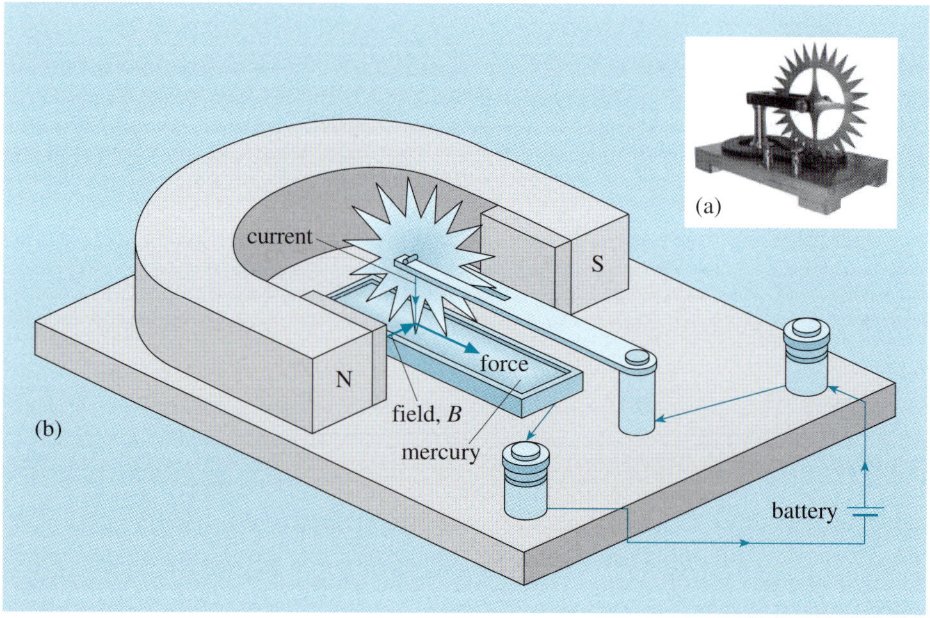

Figure 3.18 Schematic diagram of Barlow's wheel, and (inset) a model of the apparatus

---

SAQ 3.3 (Outcome 3.1)

Carefully examine Figure 3.18(b) and trace the flow of electric current through the device. If necessary, use Figure 3.12 to help your understanding of why the wheel rotates.

Insert these words and phrases in the correct order into the following passage to complete an explanation of the operation of Barlow's wheel:

current flow	inertia	positive	negative
anticlockwise	magnetic flux	force	downwards

'Conventional electric current flows from the __________ terminal of the battery supply and enters the wheel at its axle. The current then travels vertically __________ through one of the spokes towards the bath of mercury. The __________ between the poles of the horseshoe magnet is at right angles to the direction of __________. Hence the conducting spoke of the star-shaped wheel experiences a __________ that pushes it outwards, causing the wheel to move __________. As the wheel turns the current flows in successive vertical spokes, and the __________ of the wheel helps it to rotate continuously. The current flows out from the tip of the conducting spoke, through the mercury and back to the __________ terminal of the battery, completing the circuit.'

4.2 Improving Barlow's prototype

Barlow's wheel demonstrates that the problem of getting continuous power using the motor effect can be overcome by incorporating rotary motion into the design. However, there are a couple of major drawbacks that make Barlow's prototype impractical for general use. Firstly, the star-shaped conducting wheel doesn't constrain the flow of electric current in the same way that a thin wire does. This leads to non-uniform current flow and the creation of *eddy currents* that can act to oppose the motion we are trying to create. Eddy currents can have a highly detrimental effect on the *efficiency* of a motor, that is, how well it can convert electrical energy to useful mechanical energy, something that I will discuss in more detail later. The main drawback, however, is the incorporation of the bath of mercury in an attempt to ensure continuous electric contact as the wheel rotates. Clearly this doesn't make the motor very portable or safe.

In principle, these two problems can be overcome by using a *loop* of wire with *brush* contacts as shown in Figure 3.19. The brush contacts slide against *slip rings*, which allow the wire loop to rotate freely about its axis.

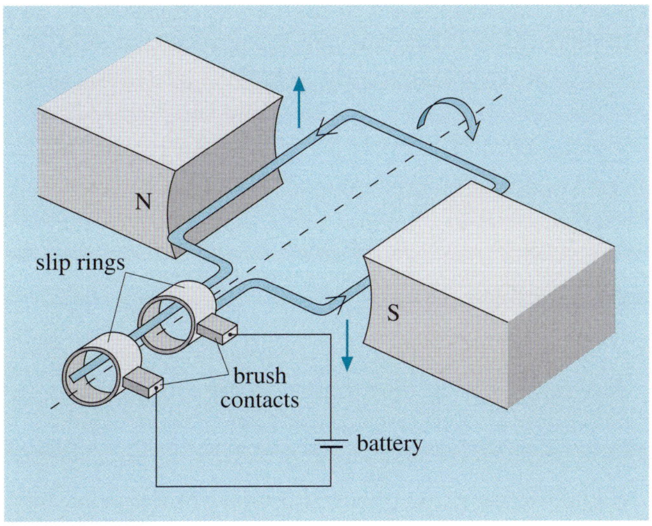

Figure 3.19 Brushes and a single coil

Now consider carefully what will happen in Figure 3.19. The loop carries electric current that flows from the positive terminal of the battery to the negative terminal. Hence the two long sides of the loop carry electric current in opposite directions in relation to the magnet, and hence experience forces in opposite directions – one up, one down. Just like a compass needle in a magnetic field, the loop experiences a torque about its axis that makes it rotate, clockwise in this case. Note that the short ends of the loop that are parallel to the magnetic field experience no force (recall $F = BIl \sin \theta$, Equation 3.2). But what happens when the coil reaches the top of its rotation? The two forces then line up with each other in opposite directions – there is no torque; no turning moment. The momentum of the coil will carry it a little way past this point but the forces will pull it back again. The coil stops moving! What is really needed is some way of switching the current round just at this point, so that the forces are suddenly reversed and the coil can continue spinning. This is achieved, quite ingeniously, using a single split-ring contact called a *commutator* as shown in Figure 3.20. It's actually a ring that is cut into two halves, each half being connected to one end of the loop.

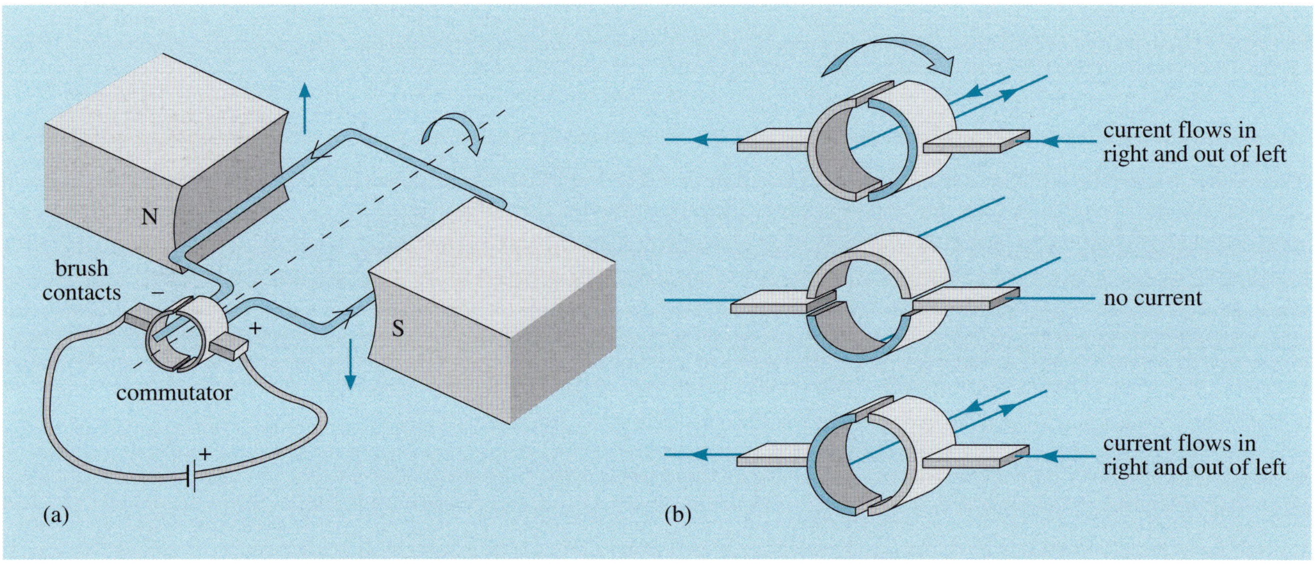

Figure 3.20 Commutator and a single coil

This time, when the coil spins round so that it is vertical the current stops flowing momentarily. However, the coil momentum carries it past this point so that the commutator and the coil turn over and the force on each half of the coil is reversed, allowing the motion to continue. Hence, in Figure 3.20, the force on the length of coil nearest the north magnetic pole is always *upwards*, and the loop rotates clockwise.

4.3 Inside a real motor

We have seen how a force can be created using the interaction between two magnetic fields, and how this force can be constrained to create rotational motion. The last section described the basic principles on which all direct current electric motors work and Figure 3.20 presented schematic pictures designed to help you understand these principles – but what do the insides of a *real* motor look like? The easiest way to find out is to look inside one for yourself. To help, I've taken apart a small motor that I bought from a mail-order electronic component suppliers for about £2. It's similar to the one you met in Block 3 Part 1, when examining the insides of an electric toothbrush, and can be powered from one or two 1.5 V batteries. Most small motor designs are quite similar. If you have trouble getting hold of one take a look inside an old battery-driven motorized child's toy – a small car or train will do. Or you could look inside an electric toothbrush, of course. If you can't find a suitable motor, or just don't want to take one apart, don't worry. You can still follow the description of the insides of my motor over the next few paragraphs by referring to Figure 3.21.

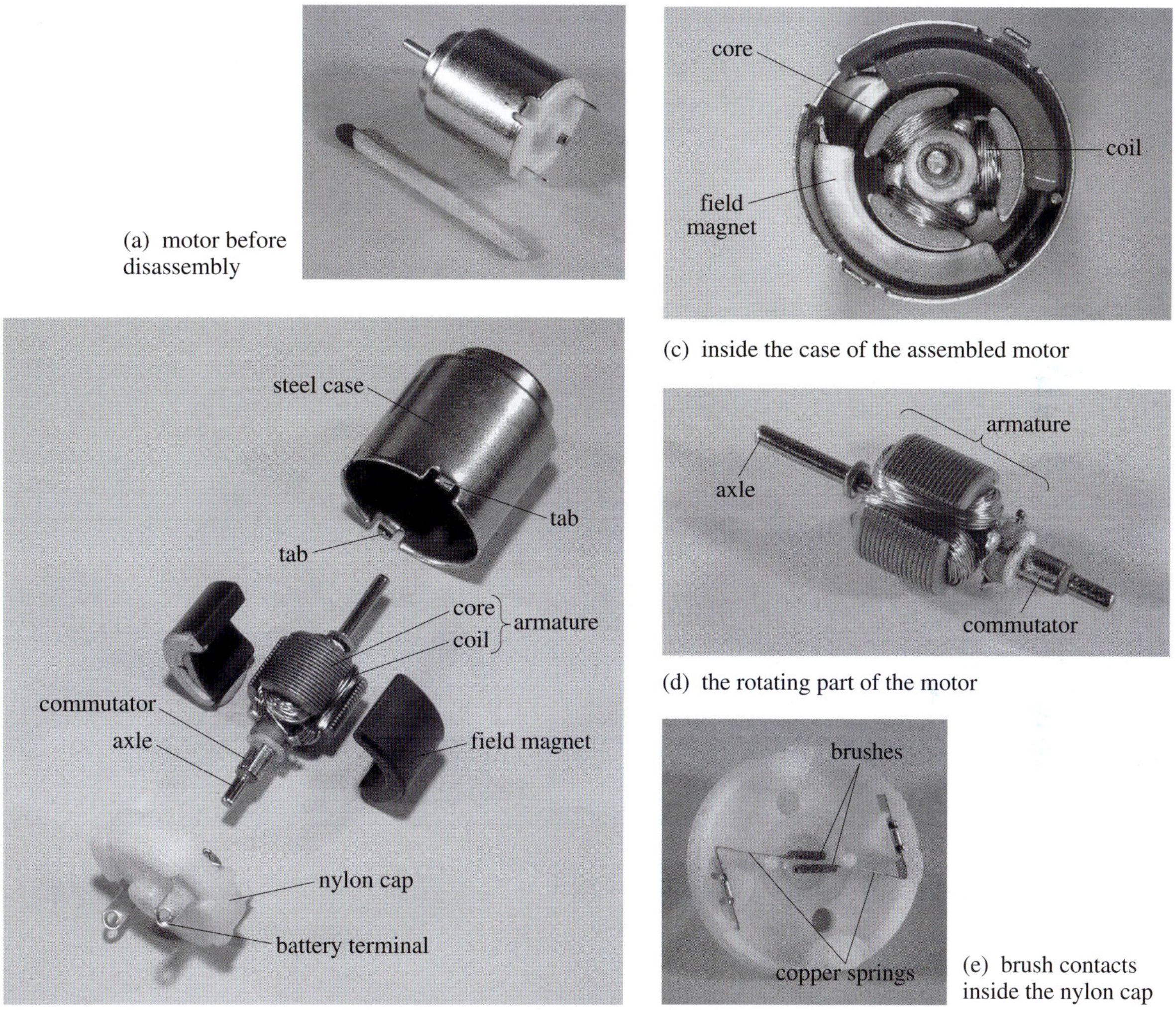

(a) motor before disassembly

(b) main components of a small d.c. permanent-magnet motor

(c) inside the case of the assembled motor

(d) the rotating part of the motor

(e) brush contacts inside the nylon cap

Figure 3.21 Components of a small d.c. motor

If you are going to take a motor apart, first take a look at its outside – it should look something like the one shown in Figure 3.21(a). In fact, my particular motor has an outer steel case, about 20 mm in diameter and 25 mm in length, with a white nylon cap at one end. The cap is held in place by two small tabs that are part of the outer casing. An axle runs through the centre and two metal connectors poke out through the nylon cap. These are the terminals for connecting to the power source – if you pulled the motor straight out of a toy or appliance they will probably have two wires attached to them that lead straight to the positive and negative terminals of the battery holder. (I tried running the motor by connecting it straight to a 1.5 V battery.

When I swapped the connections round the direction of rotation reversed. The axle spun freely at a very high rate – many times per second so far as I could tell.)

To open up the motor, bend back the holding tabs and you should be able to pull off the nylon cap. Tug on the axle and it should come out quite easily to reveal that there are many turns of copper wire wound around a metal core near its centre; this part is called the *armature*. Finally, reach inside the steel case with your finger to work free two magnets that are probably held in place by metal clips; use a screwdriver to prise them out if necessary. That's all there is to it.

Now let's examine the important components in more detail:

Permanent magnets – these provide the uniform static magnetic field with which the moving magnetic field of the armature can interact; they are usually called the *field magnets*. They are made of a *hard* magnetic material that is slow to lose its magnetism with time: see **▼Magnetic materials▲**. Note that the permanent magnets are curved to create a radial field at right angles to the plane of the rotating armature coils; this maximizes the force created.

Armature – this is the rotating electromagnetic section of the motor. The armature does the same job as the single rotating coil in Figure 3.20(a) but there are three major design improvements.

Firstly, there are a large number of turns of wire because this creates more force. In fact, having N identical current-carrying loops would be expected to increase the total torque produced N-fold. Note that the coils appear to be made of shiny, bare copper wire, but they're not. The copper wire is coated with a transparent enamel for electrical insulation. Without the insulation, the current wouldn't be constrained to each individual loop – effectively meaning that there would be only one loop instead of many.

Secondly, the turns of wire are not wound in a single coil but, in this case, *three* separate coils. These are arranged such that the planes in which they are wound intersect at 60°, forming a triangle (see Figure 3.21c). This has the advantage of allowing more turns of wire to be incorporated in a small space, and also ensures that more coil can remain within the radial field of the permanent magnets at any one time. Hence the motor is more powerful and runs more smoothly. Notice also that each set of coil windings shown in Figure 3.21(c) is *not centred on the axis of rotation*, unlike the coil we examined in Figure 3.20. In fact, the way in which the current travels in the real motor coils is a little more complicated. However, the important thing to note in Figure 3.21(c) is that the current flow in the long arms of the armature coils is mainly *downwards* near one of the magnets and *upwards* near the other. Hence the sense of the force generated is different on each side of the axis of rotation and a torque exists, as before.

Thirdly, the three coils are each wound around an iron *core*. You should be able to see from Figure 3.21 that the cores are *laminated*. This is to reduce the effect of eddy currents that are induced in them: see **▼Electromagnet induction and eddy currents▲**. Magnetically *soft* iron is used for the cores because it easily gains and loses its magnetism when the current changes direction or is switched on and off. The crystal structure of iron acts to concentrate and enhance the magnetic field passing through it, thus helping to generate a greater force. The mass of the core also provides inertia, which helps to maintain a steady angular speed.

Commutator – this switches the current between the armature coils. The commutator in Figure 3.21(d) is insulated from the axle and split into three sections. Each of the coils is connected to two different parts of the three-section commutator.

Brushes – these comprise two spring-loaded copper contacts (Figure 3.21e) that press against the commutator as the axle rotates. The brushes are in contact with two sections of the commutator at a time.

▼Magnetic materials▲

As you have worked through this part of the block you may have concluded that magnetic fields appear to come from two different sources. Firstly, there are fields arising from what I have termed 'permanent' magnets, e.g. bar and horseshoe-shaped magnets, that are usually made of an iron-bearing material. Secondly, a magnetic field is created when electric current flows through a wire. In actual fact the fundamental cause of magnetism is the same in both cases: *all magnetic fields arise from the motion of electric charges*. In the case of electric current, that's simply the flow of negatively charged electrons. In permanent magnetic materials it's the motion and interaction of the charges within each atom that creates the field rather than any large-scale flow.

When *any* substance is placed in a magnetic field it becomes magnetized to some extent, i.e. it develops a magnetic field of its own, thus enhancing the overall magnetic field. For most substances this effect is extremely weak, but for a few, notably those containing significant amounts of iron, nickel or cobalt, even a weak magnetic field causes them to become strongly magnetized – these are called *ferromagnetic* materials, after iron (*ferrum* in Latin), the commonest material of this type.

Some magnetic materials, called magnetically *soft*, are easy to magnetize and also lose their magnetism easily. Such materials, e.g. pure iron or silicon steel, are useful in electrical relays and in the core of motor armatures. Magnetically *hard* materials, on the other hand, are difficult to magnetize and do not lose their magnetism easily – these materials are used to make permanent magnets. Examples include naturally occurring lodestone (iron oxide), cobalt steel, and various aluminium–nickel–cobalt (AlNiCo) alloys.

▼Electromagnetic induction and eddy currents▲

We have seen how the combination of an electric current and a magnetic field can be used to create motion. In fact, this also works the other way round: relative motion between a conductor and a magnetic field can be used to create, or *induce*, an electric current – a phenomenon called *electromagnetic induction*. In Figure 3.22 a conducting wire is connected up in a circuit to a sensitive current-measuring device called a galvanometer. If the wire is moved upwards a current flows as shown; if the wire is moved downwards a current flows the opposite way.

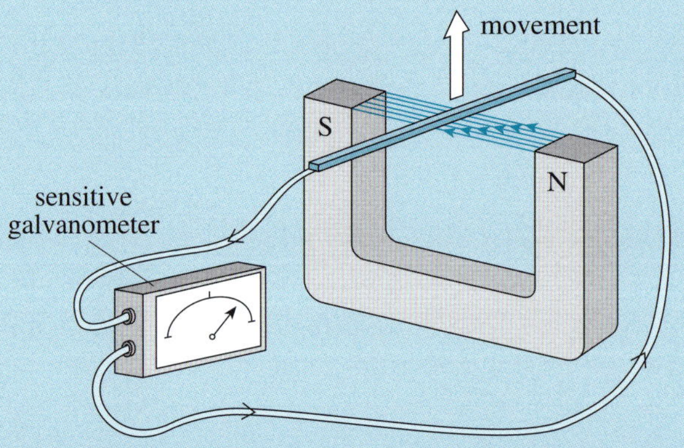

Figure 3.22 Demonstration of electromagnetic induction

Electrical engineers like to say that an *electromotive* (i.e. 'electron-moving') *force* or *e.m.f.* has been created to drive the current through the wire. The e.m.f. is sometimes called 'voltage' because it is measured in volts, but in electromagnetism the term e.m.f. is preferred. When the wire stops moving, the current also stops. The important thing here is that current flows only when the wire *cuts across the lines of magnetic flux*. If the wire is moved faster, then the rate at which the lines of flux are crossed increases, and a larger current is induced.

The rotating core of a motor armature is a conductor that cuts across the magnetic flux of the field magnets. Although the core is a large lump of metal rather than a loop of wire, localized currents, called *eddy currents*, are still induced in it. These currents don't do anything useful; they just convert electrical energy to heat energy, and so eddy currents are a nuisance and need to be reduced. The most common way of doing this is to use a laminated iron block, sliced into thin sheets that are electrically insulated from each other by a layer of varnish or an oxide coating. The orientation of the laminations is important: they must be parallel to the direction of the magnetic field (check this is true by looking at the laminations of the core in Figure 3.21b).

Exercise 3.7

Why are the laminations in an armature core aligned so that they remain parallel to the radial field lines as the core rotates?

4.4 Quantifying what a motor can do

Usually you buy a motor because you want to use it to do some work – lift a load, turn a wheel, drive a pump – not because you want to take it to pieces. To help you decide which motor might be suitable for the job, manufacturers usually supply some 'technical specifications' as a guide to motor performance. Generally, the smaller and cheaper the motor, the less information is provided. Here's what I got with my little motor:

Table 3.1 Suppliers' technical specification for a small d.c. motor

Nominal voltage / V	1.5	3.0
No-load speed / rpm	8200	14 400
No-load current / A	0.19	0.23
Maximum efficiency speed / rpm	6250	11 300
Maximum efficiency current / A	0.62	0.96
Maximum efficiency torque / g cm	6.4	10.4

The voltage range over which the suppliers advise that the motor should be used is 1.5–3.0 V. They have given a set of data for each end of this range, assuming that the user is most likely to run the motor off one or two 1.5 V batteries. The data for each voltage is of three types:

- Rotational speed, given in revolutions per minute (rpm), probably measured using an electronic tachometer. This device counts the axle rotations either remotely, using a stroboscope or optical sensor, or directly, using a small wheel or probe in contact with the axle.
- Electric current in amps. Measured simply by placing an ammeter in series with the motor and battery.
- Torque, given in 'g cm' where g means 'gram' (not to be confused with italicized g which represents 'acceleration due to gravity'). Probably recorded using a specialized electronic torque measurement transducer clamped to the axle.

You will also notice that the data is given under two operating conditions, 'no load' and 'maximum efficiency'. Like all machines, motors convert energy from one form to another. Because energy cannot be created or destroyed, the amount of energy we put into a machine must equal the total energy that comes out. You will know this as the *principle of conservation of energy*. However, it is a good idea to examine the rate at which *useful* energy is generated. In the case of electric motors this is the *mechanical* power that is made available to do work, not the *thermal* power which is lost as frictional heating in moving parts, and as electrical heating in the rotating coil, for example. The ratio of useful power output, P_{out}, to total power input, P_{in}, is called the efficiency, η (Greek letter *eta*), of a machine. Now, the 'no load' data in the table were collected while the motor was simply allowed to spin freely; no useful mechanical work was being done. On the other hand, when the motor *does* do useful work, the energy it wastes varies with the amount of work done. The suppliers have chosen to give a set of data at 'maximum efficiency' when the least proportion of energy is wasted – the optimum condition under which the motor should be run.

Note that the motor technical specifications don't actually tell us what the maximum efficiency is – have the suppliers got something to hide? In the following exercises you can correct an error in the data table and compute the efficiency for yourself.

Exercise 3.8

From your previous understanding of the term 'torque' (refer back to Block 3 Part 2, if necessary) it should be obvious that the units given in the above table for this quantity are dimensionally incorrect. Why? What should the correct SI units be? What is the conversion factor between these incorrect and correct units? (Assume $g = 9.8\ \mathrm{m\ s^{-2}}$)

Example

Using the 'maximum efficiency' data given in Table 3.1, calculate what the maximum efficiency of the motor actually is.

Solution

We can proceed using the following steps:

1. calculate the electrical power input, P_{in}:

   $$\text{electrical power} = \text{current} \times \text{voltage} = IV$$

2. convert the rotational speed, ω, to SI units (rad s^{-1})
3. convert the torque, Γ, to SI units (N m)
4. calculate the mechanical power output, P_{out}, for a rotating shaft:

   $$\text{mechanical power} = \text{torque} \times \text{angular speed} = \Gamma\omega$$

5. compute the efficiency $\eta = P_{out}/P_{in}$

The calculations for the motor running at 1.5 V are given in Table 3.2 below.

Table 3.2 Calculating maximum efficiency of the motor

Nominal voltage	1.5 V	3.0 V
Electrical power input / W	$IV = 0.62 \times 1.5$ $= 0.930$	
Angular speed / rad s^{-1}	$\omega = \frac{6250}{60} \times 2\pi$ $= 654$	
Torque / N m	$\Gamma = 6.4 \times 9.8 \times 10^{-5}$ $= 6.27 \times 10^{-4}$	
Mechanical power output / W	$\Gamma\omega = 6.27 \times 10^{-4} \times 654$ $= 0.410$	
Maximum efficiency	$\frac{0.410}{0.930} = 0.441$	

SAQ 3.4 (Outcomes 3.2, 3.3 and 3.8)

Calculate the motor efficiency when running at 3.0 V by completing the third column of Table 3.2.

From the previous example and your answer to SAQ 3.4 you will notice that more than half the power provided to the motor is wasted. You might find this a little surprising, but don't forget that this is a small, cheap motor. We will shortly consider ways to improve motor efficiency, and you will soon realize that this must come at a cost: both in terms of size and money. However, we still haven't finished with our little motor yet. We have only calculated the optimum efficiency – the best that the motor can do. In fact, the suppliers have provided enough information in Table 3.1 for us to estimate the motor response over its entire recommended operating range, as we shall see in the next two sections.

4.5 Torque–speed characteristic

A motor will generally run at a speed at which the torque generated by the electromagnetic forces matches the demand of the job – provided that the task is not beyond the motor's capability. In general, increasing the torque that is applied as a load to the motor shaft causes it to rotate at a slower rate. In fact, to a good approximation, there is a *linear* relationship between the torque applied to a d.c. permanent magnet motor and the rotational speed of its output shaft. This linearity can be represented by the straight line plotted on the graph shown in Figure 3.23.

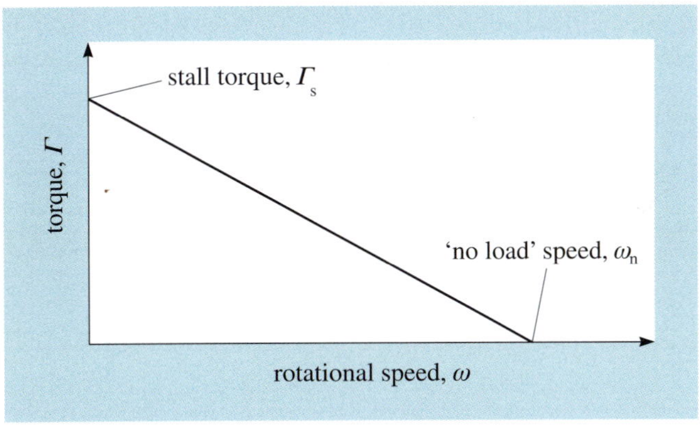

Figure 3.23 Torque–speed curve

Note that I have drawn the line such that it doesn't pass through the origin, but intercepts the torque and rotational speed axes at finite positive values. This is because:

- with no torque, the motor can spin freely at maximum speed i.e. $\Gamma = 0$ at the no-load speed, ω_n, and
- a large enough torque will cause the motor to stop spinning or 'stall' i.e. $\omega = 0$ at the stall torque, Γ_s.

There is enough information here for us to write down an equation for the torque–speed curve based on the general straight-line equation:

$$y = mx + c$$

in which x and y are variables, m is the gradient and c is the intercept on the y-axis. In Figure 3.23, ω and Γ are the variables, while ω_n and Γ_s are constants that define the gradient ($-\Gamma_s/\omega_n$) and intercepts. Hence,

$$\Gamma = -\frac{\Gamma_s}{\omega_n}\omega + \Gamma_s$$

or

$$\Gamma = \Gamma_s - \frac{\Gamma_s}{\omega_n}\omega \qquad (3.3)$$

which is the torque–speed characteristic for a d.c. permanent magnet motor. Now, check back on Table 3.1 – do you have sufficient data to plot this curve? You will notice that you are not given the stall torque, but you *do* know the torque at two speeds so you can still calculate the gradient. Follow the next example and then try it out for yourself.

Example

Using the data given in Table 3.1, and that assembled in Table 3.2, compute the gradient of the torque–speed curve for my d.c. motor running at 1.5V, and hence deduce the stall torque at this voltage.

Solution

From Table 3.2, for maximum efficiency at 1.5 V we have:

$$\omega = 654 \text{ rad s}^{-1}$$

when $\Gamma = 6.27 \times 10^{-4}$ N m

and from Table 3.1, at no load and 1.5 V:

$$\omega = \omega_n = \frac{8200 \times 2\pi}{60} = 858 \text{ rad s}^{-1}$$

when $\Gamma = 0$.

Hence the gradient:

$$\frac{d\Gamma}{d\omega} = \frac{0 - 6.27 \times 10^{-4}}{858 - 654} = -3.07 \times 10^{-6} \quad \left[= -\frac{\Gamma_s}{\omega_n} \right] \quad \text{N m s rad}^{-1}$$

and stall torque:

$$\Gamma_s = 3.07 \times 10^{-6} \omega_n = 3.07 \times 10^{-6} \times 858 = 2.63 \times 10^{-3} \text{ N m}$$

SAQ 3.5 (Outcomes 3.4 and 3.8)

Following the previous example, compute the stall torque for the motor running at 3.0 V.

SAQ 3.6 (Outcome 3.4)

Using the axes intercepts you computed in the previous example and SAQ 3.5, plot the linear torque–speed characteristics for the motor running at 1.5 V and 3.0 V on a graph of ω versus Γ. (Use the grid provided in Figure 3.24 for your plot).

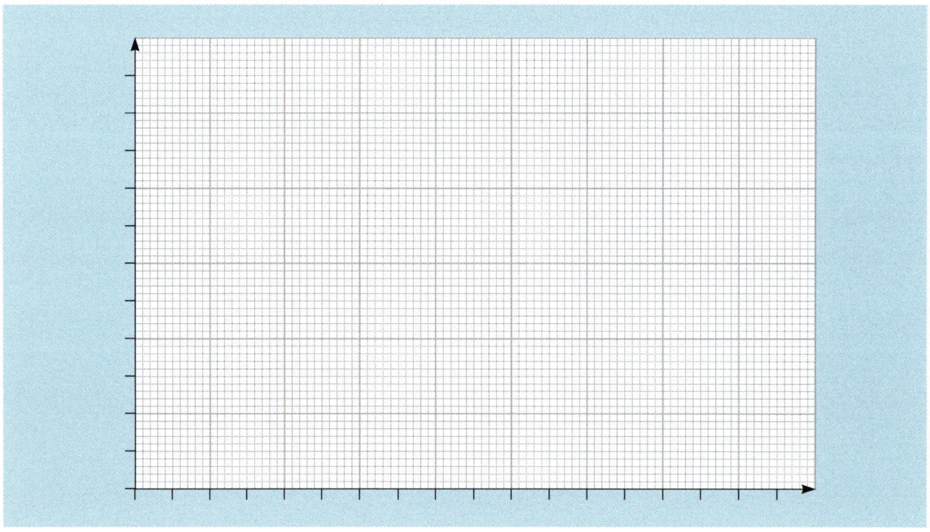

Figure 3.24

Note that the torque–speed curves we derived during the last few examples and exercises have the same slope at the two different voltages. In fact we can consider these as part of a family of curves as shown in Figure 3.25. The torque–speed curves move up and to the right with increasing input voltage.

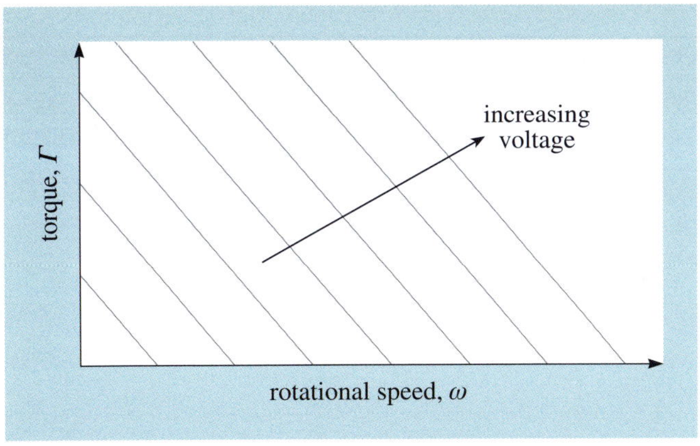

Figure 3.25 Family of torque–speed curves

4.6 Power–speed characteristic

Since the mechanical power (P) converted by the rotating motor shaft is given by torque × angular speed ($\Gamma\omega$), we can use torque–speed curves to help us understand how power varies with rotational speed. Note that power $\Gamma\omega$ corresponds to the area of a rectangle under the curve for any particular torque–speed condition, as illustrated in Figure 3.26.

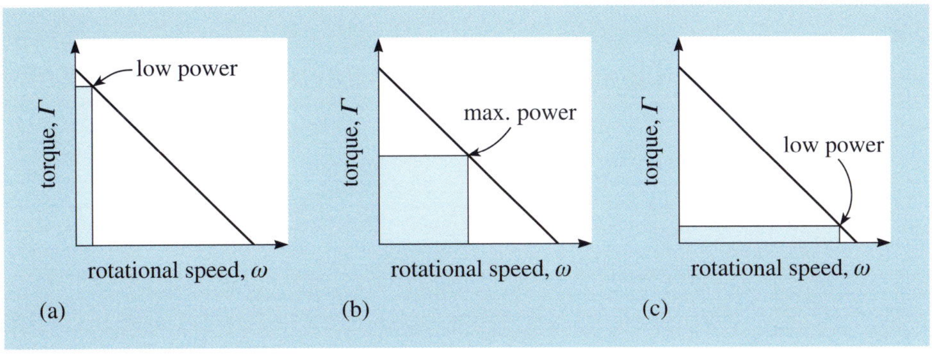

Figure 3.26 Power in relation to torque–speed

The output power is zero when the motor stalls, at $\omega = 0$, and also when it spins under no load, at $\omega = \omega_n$. Somewhere in between there is a maximum power output. It appears to be halfway. Let's prove that is the case.

Actually, quantifying the power–speed relationship is a simple matter. If we take the torque–speed Equation (3.3):

$$\Gamma = \Gamma_s - \frac{\Gamma_s}{\omega_n}\omega$$

we can just multiply both sides by ω:

$$\Gamma\omega = \Gamma_s\omega - \frac{\Gamma_s}{\omega_n}\omega^2$$

Now, since mechanical power $P = \Gamma\omega$, we have:

$$P = \Gamma_s\omega - \frac{\Gamma_s}{\omega_n}\omega^2 \qquad (3.4)$$

This is a non-linear equation because it contains a term in which one of the variables is raised to an exponent other than 1. In fact, where the independent variable (ω in this case) is squared, we call the equation *quadratic*. I've sketched this equation in Figure 3.27 and you can see that maximum power seems to occur at about $\omega_n/2$.

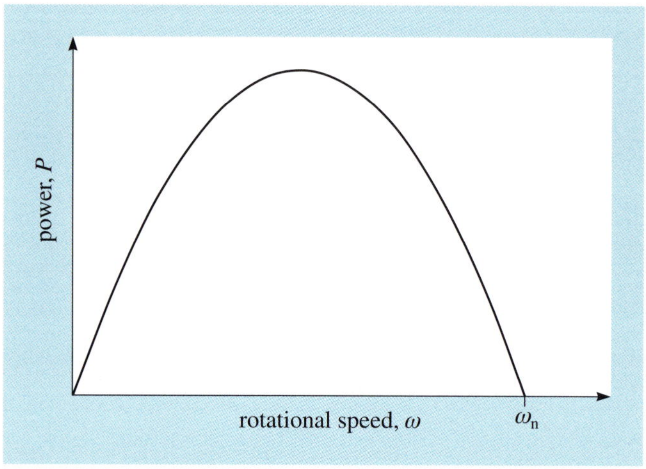

Figure 3.27 Power–speed curve

To prove that this is *exactly* true we can use a little calculus. The maximum power occurs at the 'top' of the curve, when its gradient is zero, i.e. when

$$\frac{dP}{d\omega} = 0$$

and differentiating the power–speed Equation (3.4) we get:

$$\frac{dP}{d\omega} = \Gamma_s - 2\frac{\Gamma_s}{\omega_n}\omega = 0$$

So at maximum power,

$$\Gamma_s = 2\frac{\Gamma_s}{\omega_n}\omega$$

or

$$\omega = \tfrac{1}{2}\omega_n \qquad (3.5)$$

which proves we were right all along!

If we substitute Equation (3.5) back into the power–speed Equation (3.4) we get an expression for P_{max}, the maximum power:

$$P_{max} = \tfrac{1}{2}\omega_n\Gamma_s - \frac{\Gamma_s}{\omega_n}\left(\tfrac{1}{2}\omega_n\right)^2 = \tfrac{1}{2}\omega_n\Gamma_s - \tfrac{1}{4}\omega_n\Gamma_s$$

$$P_{max} = \tfrac{1}{4}\omega_n\Gamma_s \qquad (3.6)$$

SAQ 3.7 (Outcomes 3.5 and 3.8)

Using the values for ω_n and Γ_s you calculated in SAQ 3.5, compute the maximum mechanical power of my small motor when running at 3.0 V.

The answer to SAQ 3.7 tells us that the maximum mechanical power the motor can generate is 1.76 W. But if you look back at your answer to SAQ 3.4 (Table 3.2) you will see that the power at which the motor runs most efficiently is 1.20 W. In other words, it seems that the requirements for getting the most out of the motor are different from those for using it at optimum efficiency. I'll come back to this trade-off between power and efficiency when we consider motor design in a little more detail later.

4.7 Lifting and shifting

I've spent some time showing you how to quantify the various characteristics of my d.c. motor. I've also pointed out that the torque–speed–power relationships are important in helping an engineer decide what a motor can do, and how well it can do it. But how exactly does the decision-making process work in practice?

As an example I'll take the simple case of using my motor to lift a load. I've got a small Penguin paperback book on my desk (*Metals in the Service of Man* by Street and Alexander – an excellent introduction to metallurgy, in case you're interested!), I wonder if it can lift that? Unfortunately, I have already taken my motor to pieces, so I can't actually try it – I'm forced to use the characteristics we've already deduced. But, as I've said, if I were a design engineer trying to develop a book-lifting device for lazy librarians it would make sense to figure out if the motor can do the job first, before actually going out to buy it.

Imagine the motor is fixed to a desktop and its shaft is connected to a spindle, diameter 2 cm, as shown in Figure 3.28. String attached to the spindle is also tied to the 250 g book lying on the floor 1 m below.

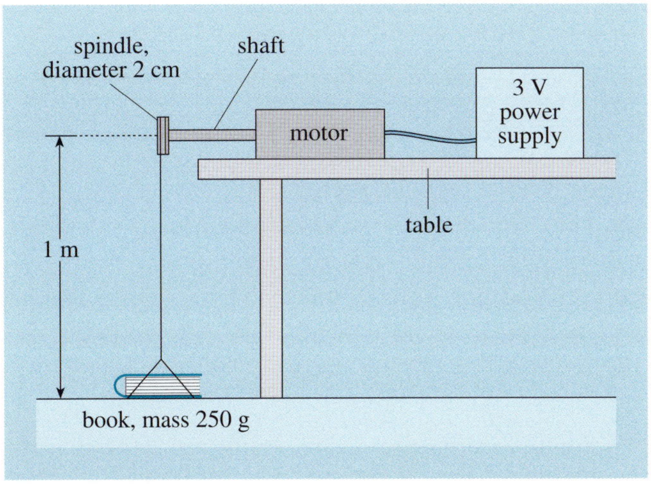

Figure 3.28

The idea is that when the 3 V supply to the motor is switched on, the spindle turns and winds up the string around its circumference, lifting the book from the floor. So, is the little motor (mass 25 g) capable of lifting something ten times its own weight, and, if so, how quickly will it do it? Let's see.

Example

Consider using my small d.c. motor to lift a book by means of the arrangement shown in Figure 3.28, described above.

The motor is driven from a 3 V supply and has stall torque $\Gamma_s = 4.67 \times 10^{-3}$ N m and no-load speed $\omega_n = 1510$ rad s^{-1}, just as you computed in your answer to SAQ 3.5.

Can the motor lift the book? If so, how fast?

Solution

If the torque necessary to lift the book is less than the motor's stall torque, then it ought to be able to do the job.

Assume that the string is weightless and of negligible thickness, and ignore any friction effects. Also disregard the fact that it is necessary to accelerate the book from rest to get it moving (this would require more force).

Note that the mass, *m*, of the book exerts a force, *mg*, that acts downwards at the outer radius, *r*, of the spindle, 1 cm from its axis of rotation. Hence, using SI units:

$$\begin{aligned}\text{torque required to lift book} &= \text{turning force} \times \text{distance from axis of rotation}\\ &= (0.25 \times 9.8) \times 0.01\\ &= 2.45 \times 10^{-2}\ \text{N m}\end{aligned}$$

This is much larger than Γ_s you computed in SAQ 3.5, so the motor would stall and fail to lift the book.

Apparently the motor just isn't up to it. I think I'll buy another one. But before I do it would be interesting to know just how much it *can* lift in this way.

Exercise 3.9

Consider again the book-lifting arrangement in Figure 3.28. If the motor stall torque is $\Gamma_s = 4.67 \times 10^{-3}$ N m as before, what is the largest mass it can lift using this set-up? (Assume $g = 9.8$ m s^{-2})

The motor can lift about twice its own weight, just under 0.5 N. But something is not quite right here. My three-year-old daughter has a wooden train set, and I know that the engine contains one of these motors. Yet that motor can haul itself, three carriages and various pieces of track (total mass about 250 g) noisily up and over bridges and viaducts for hours on end. Another thing: the shaft on my motor spins like crazy, 11 300 rpm when used at optimum efficiency (Table 3.1), but the engine's wheels grind around at about 120 rpm. The vital addition to the toy engine's internal workings is, of course, a gearing mechanism that changes the speed of rotation. The same output power can be obtained from a shaft by running at high speed with low torque, or by running at low speed with high torque (you covered this in Part 2 of this Block). Hence, before looking at exactly how gears might work in this case, we really ought to see if our motor has enough *power* to do the job. To do this I need to decide how quickly the book should be lifted. I'll say 5 seconds to cover 1 m, which ought to be slow enough for me to react in time and turn the motor off when the book reaches a satisfactory height.

Exercise 3.10

(a) What power is required to lift a 250 g book over a distance of 1 m in 5 s? (Assume $g = 9.8$ m s^{-2})

(b) Is the motor capable of supplying this power?

The motor *is* capable of lifting the book. I just need to reduce the speed of rotation of the spindle, and hence increase the torque it can generate, by using an appropriate ▼**Gear ratio**▲. A set of miniature gears is available in kit form from the same suppliers that provided me with the motor. From the kit I can assemble gear trains that will give me ratios of 4:1, 16:1, 64:1 or 256:1 – I wonder if they are of any use? The next example and SAQ will take you through a gear selection process.

Example

Consider, again, using my small d.c. motor to lift a book, this time by means of the spindle (diameter, 2 cm) connected via a gearbox to the shaft of the motor, as shown in Figure 3.29.

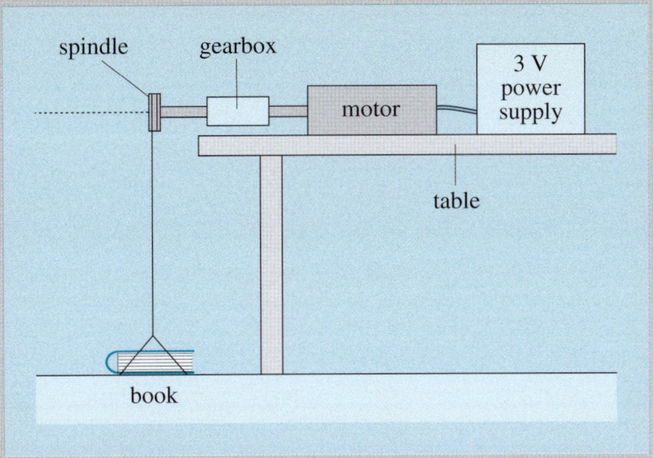

Figure 3.29

The motor is driven from a 3 V supply and has stall torque $\Gamma_s = 4.67 \times 10^{-3}$ N m and no-load speed $\omega_n = 1510$ rad s^{-1}. The spindle torque necessary to lift the book is 2.45×10^{-2} N m. How long will it take to lift the book through 1 m at constant speed using a 16:1 gear train to magnify the torque that can be generated by the motor shaft? (Assume no power losses in the gear train.)

Solution

Note that although the power generated by the motor shaft and by the spindle are the same, the torque and angular speed are different in each case.

1 First calculate what the torque, Γ_{motor}, generated by the motor shaft would need to be.

We know that the torque generated at the spindle to lift the book is

$$\Gamma_{spindle} = 2.45 \times 10^{-2} \text{ N m}$$

This has been magnified 16-fold by the gear train, hence the torque at the motor shaft is

$$\Gamma_{\text{motor}} = \frac{2.45\times10^{-2}}{16} = 1.53\times10^{-3}\ \text{N m}$$

2 Calculate the corresponding angular speed, ω_{motor}, of the motor shaft.

Using the torque–speed Equation (3.3) we have:

$$\Gamma_{\text{motor}} = \Gamma_{\text{s}} - \frac{\Gamma_{\text{s}}}{\omega_{\text{n}}}\omega_{\text{motor}}$$

and rearranging:

$$\frac{\Gamma_{\text{s}}\omega_{\text{motor}}}{\omega_{\text{n}}} = \Gamma_{\text{s}} - \Gamma_{\text{motor}}$$

$$\omega_{\text{motor}} = \frac{(\Gamma_{\text{s}} - \Gamma_{\text{motor}})}{\Gamma_{\text{s}}}\omega_{\text{n}}$$

$$= \frac{(4.67\times10^{-3} - 1.53\times10^{-3})}{4.67\times10^{-3}}\times1510 = 1020\ \text{rad s}^{-1}$$

3 Calculate the mechanical power.

$$P = \Gamma_{\text{motor}}\omega_{\text{motor}} = 1.53\times10^{-3}\times1020 = 1.56\ \text{W}\ \left[= \Gamma_{\text{spindle}}\omega_{\text{spindle}}\right]$$

4 Calculate the time, t, to lift the load.

The mechanical power is also the rate of work done in raising the 250 g book through 1 m, i.e.

$$P = \frac{Fd}{t}$$

$$t = \frac{Fd}{P} = \frac{(0.25\times9.8)\times1}{1.56} = 1.57\ \text{s}$$

SAQ 3.8 (Outcomes 3.3, 3.4, 3.6 and 3.8)

Use the previous example to complete the Table 3.3 for the book-lifting task of Figure 3.28, using each of the gear trains available. Which gear ratio is most suitable to achieve this task in about 5 s?

Table 3.3 Gear ratio calculations for SAQ 3.8

Gear ratio	4:1	16:1	64:1	256:1
Shaft torque Γ_{motor} / N m		1.53×10^{-3}		
Shaft speed ω_{motor} / rad s^{-1}		1020		
Power P / W		1.56		
Time for lifting through 1 m, t / s		1.57		

So, the gears available allow the book to be lifted pretty much as planned. Notice how the suitability of the 64:1 gear train emerged as a solution to the problem on taking a routine approach: gear trains known to be available were assessed for suitability (rather than inventing gear ratios).

Torque, speed and gear ratios seem to crop up everywhere. Here are a couple of problems to help ensure you have fully grasped the theory.

There is a motor-related Mathcad worksheet on the T207 CD-ROM.

SAQ 3.9 (Outcomes 3.5 and 3.7)

The power–speed curve for the ungeared motor in Figure 3.28 is plotted in Figure 3.30. The motor has stall torque $\Gamma_s = 4.67 \times 10^{-3}$ N m and a no-load speed of $\omega_n = 1510$ rad s^{-1}.

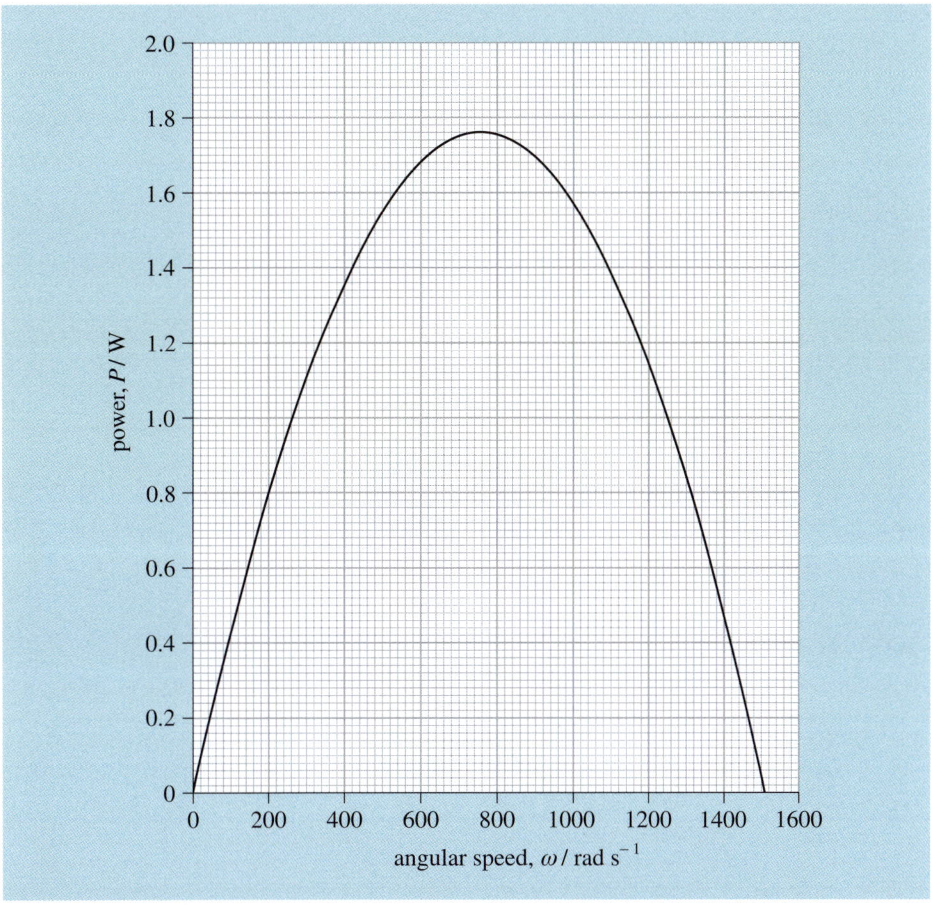

Figure 3.30

(a) From the graph, read off the values of ω that correspond to an output power of 0.49 W (you computed this power in Exercise 3.10).

(b) Read off the maximum power from the graph. Check that it corresponds with the value you computed for the motor in SAQ 3.7.

SAQ 3.10 (Outcomes 3.6, 3.7 and 3.10)

The engine of a toy train set contains a motor just like the one we have been examining and runs from a single 1.5 V cell, so assume stall torque, $\Gamma_s = 2.6 \times 10^{-3}$ N m and no-load speed $\omega_n = 860$ rad s^{-1}.

(a) When I pick the train up so that its wheels *can spin freely* they turn at around 120 rpm. What gear ratio has probably been used in the engine? (Choose from 4:1, 16:1, 64:1 or 256:1)

(b) Why did the speed ratio you calculated in (a) not coincide exactly with the gear ratio used?

(c) When the engine chugs uphill, pulling a particularly heavy load, the motor shaft spins at *three-quarters* of its no-load speed. Calculate the mechanical power that is generated by the motor under these circumstances.

▼Gear ratio▲

Gears are machines that can be used to magnify either torque or rotational speed. What they don't do is increase the amount of power available. In fact, a gearbox, like any machine, is never 100 per cent efficient and a small amount of power is always lost.

A gear consists of a circular disc or cylinder with teeth cut into it that engage with the teeth on another gear and make it rotate in the opposite direction. With gears of the same diameter and number of teeth, a single revolution of one gear causes a single revolution of the other. This is expressed as a 1:1 gear ratio. If two gears with different numbers of teeth are used, the speed of rotation of one of the gears will be faster or slower than the other. In Figure 3.31 the larger gear has twice as many teeth as the smaller one. Since the teeth always interlock the smaller wheel must have to complete two rotations for every one that the larger wheel does.

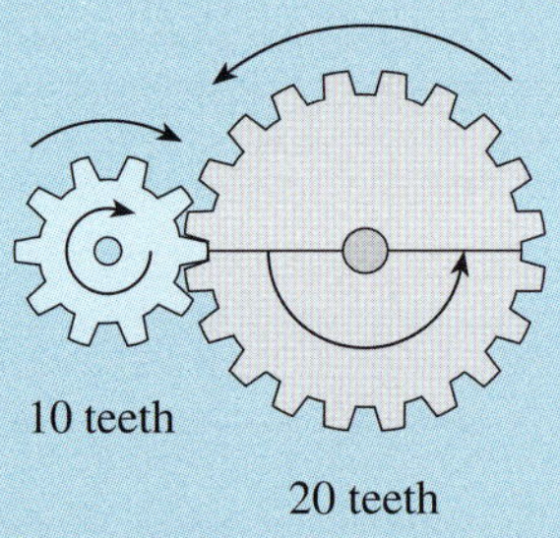

Figure 3.31 Gears with a 2:1 ratio

This might be clearer if you imagine the gears simply as rotating circles. The larger diameter circle has a bigger circumference, and hence the smaller circle must have to travel faster to keep up.

If the large wheel in Figure 3.31 is used to drive the smaller wheel then the rotational speed, ω, is magnified; in fact it is doubled. Ignoring efficiency losses. all the power, P, from the large wheel must be transferred to the small wheel. Now, since the power remains the same and $P = \Gamma\omega$, doubling the angular speed must halve the torque. Hence speed is magnified but torque is reduced. On the other hand, if the small wheel is used to drive the large wheel then torque is magnified, again in the ratio 2:1, and rotational speed is reduced in the same ratio.

Gears can come in various ratios but large gear ratios are best produced using *gear trains* in which several gears are joined together.

4.8 Optimizing motor efficiency

There's one final aspect of my little d.c. permanent magnet motor that needs to be dealt with: why does it spin so fast? Whenever you see a motor of this type in use it invariably has some kind of gearing mechanism attached in order to reduce speed and increase torque. Why not just have the motor running slowly in the first place? Well, it's all to do with efficiency.

We can start by recalling that mechanical work, $F\Delta x$, is done when a force, F, causes a small movement Δx in the direction of the force. Therefore, in time Δt the mechanical power generated is:

$$P_{\text{mech}} = \frac{F\Delta x}{\Delta t}$$

Now, our motor derives its force from several current-carrying conductors of length l so that, assuming they lie perpendicular to a radial field, we have, from Equation (3.2):

$$F = BIl\sin\theta = BIl$$

since $\theta = 90°$ and $\sin\theta = 1$. The mechanical power per conductor can therefore be rewritten as:

$$P_{\text{mech}} = BIl\frac{\Delta x}{\Delta t}$$

that is,

$$P_{\text{mech}} = BIlv \qquad (3.7)$$

where v is the linear velocity of the conductor. But the current that helps create this force also generates thermal energy through resistance heating in the armature coil. The power lost in this way can be expressed as:

$$P_{\text{therm}} = I^2R = I^2\frac{\rho l}{A} \qquad (3.8)$$

where A is the cross-sectional area of the conductor with resistivity, ρ. There are several other important sources of motor power loss. However, I want to try to isolate certain factors that influence efficiency, so for the time being I'll ignore all other power losses and just express the efficiency of our motor as:

$$\eta = \frac{P_{\text{out}}}{P_{\text{in}}} = \frac{P_{\text{out}}}{P_{\text{out}} + P_{\text{lost}}} = \frac{P_{\text{mech}}}{P_{\text{mech}} + P_{\text{therm}}} = \frac{1}{1 + \dfrac{P_{\text{therm}}}{P_{\text{mech}}}}$$

from which it is clear that the highest efficiencies will be gained by having low values of $P_{\text{therm}}/P_{\text{mech}}$. By using Equations (3.7) and (3.8) we can write:

$$\frac{P_{\text{therm}}}{P_{\text{mech}}} = I^2\frac{\rho l}{A}\frac{1}{BIlv}$$

that is,

$$\frac{P_{\text{therm}}}{P_{\text{mech}}} = \frac{\rho I}{BvA} \qquad (3.9)$$

and it is now possible to make several deductions about designing a powerful motor with efficiency in mind.

Exercise 3.11

The preceding analysis indicates that to get a large power output we need P_{mech} to be *large* (Equation 3.7), but to get high efficiency P_{therm}/P_{mech} (Equation 3.9) should be *small*.

Six different variables appear on the right-hand side of these equations: A, B, I, l, v and ρ. Imagine what would happen if each of these variables were increased in turn, while keeping all the other variables constant.

(a) Increasing which *three* of these variables would improve efficiency?

(b) Increasing which *two* of these three would *also* increase power output?

For greatest efficiency and power, permanent magnet motors ought to have a large magnetic flux density, B, and high speed, v, of the rotating part of the motor – the rotor. That is why my motor spins so fast. Note that v is a linear, not angular, velocity. The outer rim of a larger rotor will travel at a much greater linear speed than a smaller rotor with the same angular velocity (or rpm). Hence using a large diameter core and running the motor at high speed improves efficiency. In this respect, bigger is better. Of course, having a bigger motor would also allow me to incorporate a larger, stronger magnet – further increasing both efficiency and power!

There's one last aspect of efficiency to mention. Looking back at my simplified analysis of motor efficiency, it is apparent that P_{therm}/P_{mech} approaches zero as the motor is allowed to run at ever faster speeds, indicating efficiencies approaching 100 per cent. At least two of the factors that we have neglected become increasingly important at higher speeds – (1) friction in brushes and bearings; (2) windage (pronounce wind as in windmill), which is related to aerodynamic drag on the spinning rotor. The inclusion of these terms leads to an overall decline in efficiency beyond a certain speed; thus for any motor there is a speed at which the efficiency is at its highest.

SAQ 3.11 (Outcome 3.2)

Explain the following.

(a) Doubling the effective length of a motor (i.e. the total length of the rotor coil wire) doubles the power output.

(b) Efficient motor designs use the thickest wire that can be fitted into the available space.

4.9 Summary of Section 4

- A d.c. permanent magnet motor uses the forces that arise between two sets of interacting magnetic fields – one from fixed permanent magnets, the other from current-carrying rotor coils – to create circular motion.
- Because the forces in different parts of the d.c. motor armature act in different directions, it is necessary to swap round the current direction every half turn, using commutators and brushes, to ensure continuous motion.

- Motor efficiency is the ratio of mechanical power output to electrical power input.
- There is a linear relationship between torque and rotational speed for the d.c. permanent magnet motor.
- The relationship between power and rotational speed is non-linear (quadratic), with maximum power occurring at half the 'no-load' speed for this type of motor.
- Greatest efficiency and power are achieved by designing a d.c. motor to run at high speed; gears are necessary to increase torque and reduce speed for a given power.

5 MORE COMPLEX MOTORS AND GENERATORS

In Section 4, I looked in a fair amount of detail at how a very small motor works. I dismantled and studied the motor at length so that you could get an understanding of the fundamental principles that determine its characteristics. In this respect the simplicity of the motor was helpful. However, as you might imagine, things can get a little more complicated if we wish to generate greater power, to shift a large load or move a vehicle, for example. Nevertheless, the basic principles that you are now familiar with will help you to appreciate how larger, more complex electric motors work. And if you know how a motor works, you're well on the way to understanding how a generator works too.

5.1 Larger d.c. motors

Figure 3.32 shows a d.c. motor whose optimum output power is rated at two horsepower by the manufacturers. That's about 1.5 kW (one horsepower is equivalent to 746 W), so it is useful for driving machine tools, such as lathes, or small conveyor belts. Although it's about 40 cm long, weighs over 300 N and does the same job as several thousand of my little motors, I still found it in a catalogue titled 'Small Motors', so everything's relative.

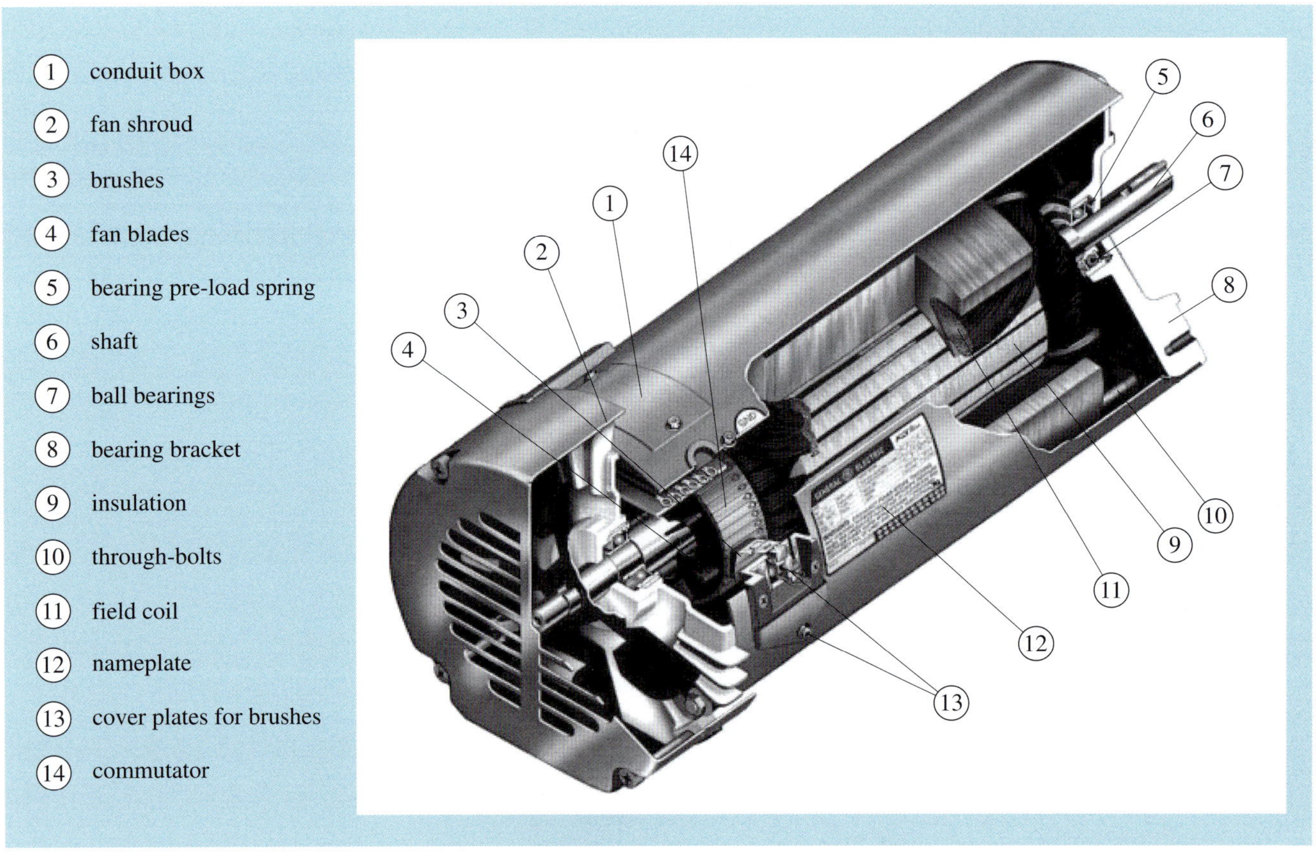

Figure 3.32 Cut-away view of a 1.5 kW d.c. motor

Exercise 3.12

Compare the 1.5 kW, 100 V motor in Figure 3.32, with the 0.4 W, 1.5 V motor in Figure 3.21. Try to list at least five significant design changes that have been implemented in the case of the larger motor.

In examining the two motors you will have noticed that one of the major design alterations involved replacing the field magnets with further coils of copper wire, collectively called the *field coil*. You know that every electric current has a magnetic field associated with it; the field associated with a straight wire is circular (Figure 3.11b), for example. Well, it just so happens that a long *cylindrical* coil of wire, called a *solenoid*, has a magnetic field that is similar to that of a bar magnet, see Figure 3.33. Hence solenoids can be used as direct replacements for the field magnets we looked at earlier.

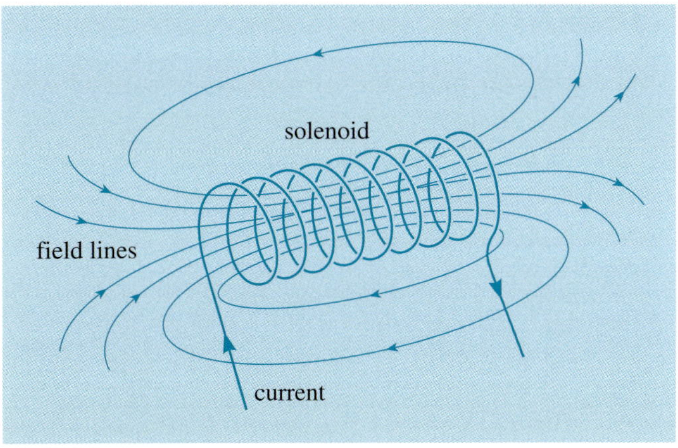

Figure 3.33 Field created by a solenoid

As you might expect, the magnetic strength of a solenoid depends on the current supplied, the number of coil turns and the nature of the core around which they are wound. This makes 'field coils' an extremely flexible addition, because their magnetic field can be easily altered to suit varied design requirements. Nevertheless, permanent magnets are often favoured for small motors, although their use is rare in motors larger than a few kW. This is partly due to the cost-effectiveness of using small permanent magnets instead of diverting current to energize the field coils. Permanent magnet motors run cooler than wound field-types and no power is used, and wasted, in maintaining the static magnetic field.

All of which brings us nicely on to a second major design change in the larger motor – the incorporation of a cooling fan. Electrical heating causes thermal energy to be generated in the armature and field coils; heat is also dissipated due to friction in rotating parts and at the contact between the brushes and commutator, for example. If the rate at which this heat is generated is greater than the rate at which the motor can lose heat to its surroundings, then the temperature of the motor will rise. Excessive temperature increases can damage the motor components, particularly the coils, and shorten the life of the motor.

Exercise 3.13

Field and armature coils are made of copper wire which has a melting point of around 1080 °C. The working temperature of the motor in Figure 3.32 shouldn't exceed 100 °C. Why is heat-induced coil damage still the limiting factor in determining the life of electric motors of this type?

So why does the big motor need cooling down when the little one doesn't? Well, how quickly an object loses heat depends on a number of factors including the size and shape of the object, the temperature difference between the object and its surroundings, and the nature of the surface of the object. So far we have concentrated on the differences between the two motors, but in many respects they are quite similar. They have roughly the same external shape, similar internal components and are made of comparable materials, mainly metals. If they were at the same temperature the main difference between them affecting the rate of heat lost to the environment would be their surface areas. Clearly, the larger motor ought to generate more heat than the smaller one – but on the other hand it's bigger so it ought to lose more. We might conclude that the increase in surface area of the larger motor hasn't kept pace with the increase in heat it needs to lose. Let's check whether that is true in the following exercise.

Exercise 3.14

Consider our two motors running at steady speed: (i) the small motor (Figure 3.21, external dimensions 25 mm long by 20 mm diameter) with a mechanical power output of 0.41 W at 44 per cent efficiency, and (ii) the larger motor (Figure 3.32, external dimensions 40 cm long by 20 cm diameter) with a mechanical power output of 1.5 kW at 70 per cent efficiency.

(a) Determine how many watts of heat are generated in each motor.

(b) Estimate the surface areas of each motor and so compare the amount of heat energy that must be lost per square metre of surface, when the temperature of the motors has stabilized.

To help attach significance to quantities of heat or power I often think in terms of electric heaters and light bulbs. The maximum power you can get from any single appliance plugged into the normal domestic UK 13 A mains supply is about 3 kW. Hence a single bar on one of those 3-bar electric heaters you may have seen generates about 1 kW of power as heat energy. Electric light bulbs are frequently rated at 60 W. In other words, the 640 W power loss in the large motor you computed for the last exercise is equivalent to over 10 light bulbs or the heat from a over half of a bar on the electric heater. No wonder a cooling fan is necessary.

A couple of the other design changes ought to be a little more obvious to you by now. The power of the motor in Figure 3.32 has clearly been enhanced over that of the smaller motor by using an armature of larger, more numerous coils rotating in a stronger magnetic field. Steel ball bearings have been incorporated to facilitate the easy rotation of the motor shaft at greater loads, thus reducing friction losses. You'll find out more about bearings of various sorts in Block 4. Somewhat less obvious is the mounting of the brushes, which are subject to a large amount of wear, on the inner surface of the frame. This means that the brushes can be changed easily without dismantling the motor, and also permits the use of longer brushes that have to be changed less often.

Finally, take a look at Figure 3.34, which shows the general design layout for a d.c. motor capable of generating up to 4000 horsepower (3 MW). That's equivalent to about two thousand of the 'large' motors we've been looking at, or 6 million of the small ones.

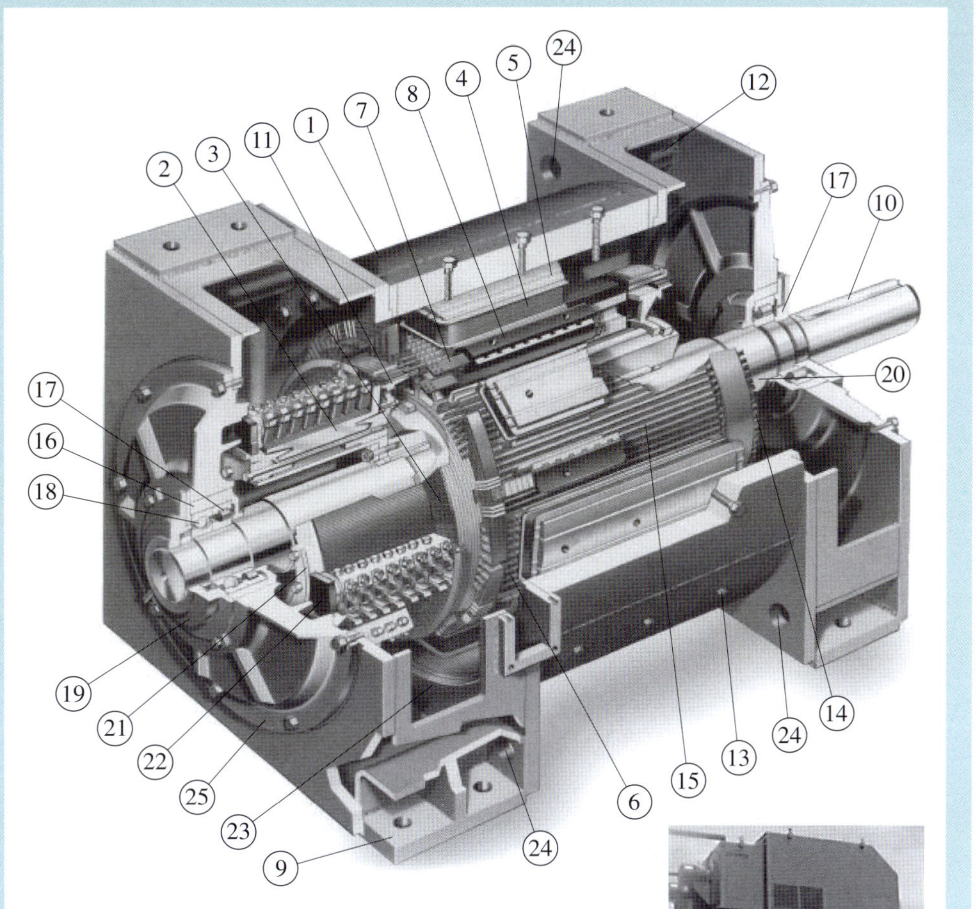

Figure 3.34 Large variable-speed d.c. motor. Inset shows external view of the motor, with a large cooling unit above

Despite this huge difference in the amount of power generated, you should still be able to identify all the key motor components – although the cooling fan is notably absent. In these cases it is often necessary to circulate water, a more effective coolant than air, around the outside of the motor. The efficiency of this huge motor is around 90 per cent. This means that there are 300 kW of wasted energy to get rid of – that's 300 single-bar electric heaters! The cooling apparatus associated with this motor (see the inset of Figure 3.34) is almost as large as the motor itself. The 90 per cent efficiency of the 3 MW motor can be compared to 70 per cent for the 1.5 kW motor and 44 per cent for the small one. This agrees with the conclusions we drew from Section 4.8 about the general influence of motor size on efficiency.

5.2 Generating electric current

We know that the combination of an electric current and a magnetic field can be used to create motion (the motor effect), and also that relative motion between a wire and a magnetic field can be used to create current (electromagnetic induction). Figure 3.35 shows a single coil of wire free to rotate in a radial magnetic field. We can create a current by physically rotating the coil so that the wire cuts across the lines of magnetic flux that make up the magnetic field. We've turned a motor into a generator!

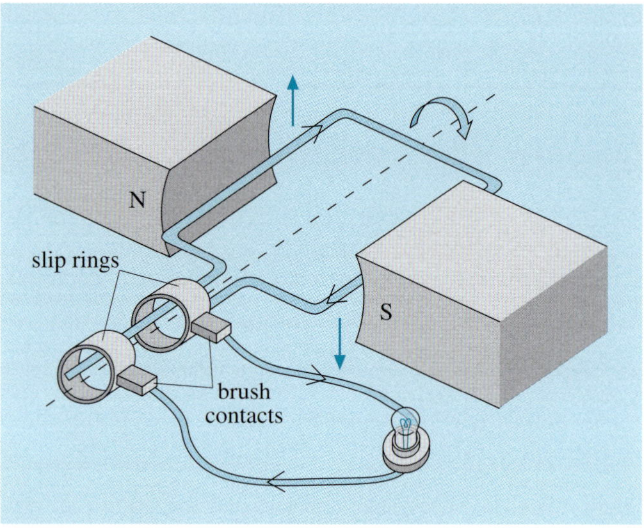

Figure 3.35 A simple generator

Now, compare Figure 3.35 to Figure 3.19 and you will see that the arrangement is almost the same, except this time the coil isn't supplied with a current from a battery. In Figure 3.19 the e.m.f. to drive the current is provided by the chemical energy in the battery. In Figure 3.35 mechanical energy is used to create what is called an *induced* e.m.f. For this type of generator, just as in the case of the motor, the direction of the induced current depends on the direction of motion of the wire in relation to the fixed magnetic field. In Figure 3.35 the left arm of the coil is indicated to be moving upwards and the electric current flows as shown, but half a turn later the same section of wire will be moving downwards and the current direction will be reversed. Hence this generator creates a varying electric current whose direction changes every half turn (Figure 3.36a). This is an *alternating current*. If a split-ring commutator were added to prevent the reversal of current just when the coil is vertical then a unidirectional but varying *direct current* can be generated (Figure 3.36b). This can be smoothed out to reduce the peaks, giving a steadier d.c., by using electronic circuitry.

Now, it may have occurred to you that if a motor and a generator are basically the same thing then a motor must *generate* current, even when it is being a motor. Well, that's correct, but it doesn't make the motor any more powerful – quite the opposite in fact. Compare again Figures 19 and 35. For the same coil position and direction of rotation, the flow of current in the motor is *opposite* to that in the generator. In other words *the e.m.f. induced in a motor opposes the current that makes the coils turn*. In a way that makes sense. If it were the other way round, i.e. the induced current further assisted the motor rotation, we'd end up with a perpetual motion machine that could keep itself running – and that would defy the principle of conservation of energy. Nevertheless, quantifying the e.m.f. induced in the

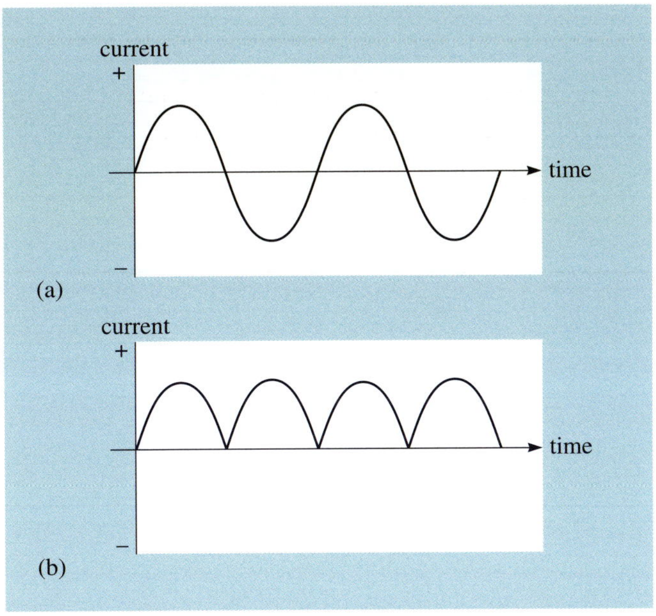

Figure 3.36 Current magnitude from a simple generator: (a) a.c., (b) d.c.

rotating coil of an electric motor gives us new insight into how the motor works and how the power is dissipated within it: see ▼**Induced e.m.f. and motor power**▲.

In modern power stations, electric generators are used to produce huge voltages and currents, up to about 25 kV and 1000 A. The generator design in Figure 3.35 can obviously be improved by adding a core and more coils to the rotating armature, and by incorporating stronger electromagnetic field coils. However, a voltage drop always occurs at the sliding brush contacts and can become a serious problem when large currents and voltages are generated. The solution, shown schematically in Figure 3.37, is to keep the armature stationary (now called the *stator*) and rotate the electromagnets (the *rotor*).

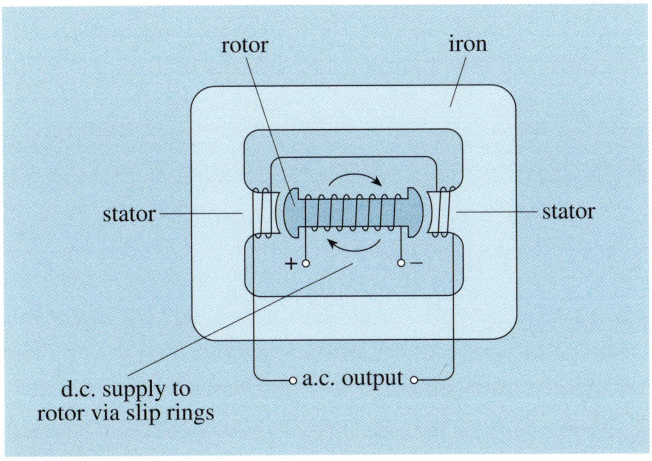

Figure 3.37 Schematic diagram of an industrial generator (alternator)

This has the advantage of allowing the large voltage and alternating current generated to be fed away directly, via fixed contacts. Only a relatively small direct current needs to be passed through slip rings to energize the field coils. In a power station, the rotor is usually driven by a water or steam turbine that also drives a small d.c. generator to power the electromagnet.

▼Induced e.m.f. and motor power▲

The rotating coil of a motor armature has an e.m.f. induced in it. This e.m.f is proportional to the speed of rotation of the armature, and it opposes the current that has been supplied using the potential difference from an external power source. If the applied potential difference is V and the induced e.m.f. is E, and ignoring any voltage drop that occurs at the brush contacts, then the total voltage difference that drives the current must be $(V - E)$. Hence using Ohm's law we can write:

$$V - E = IR$$

where I and R are the coil current and resistance respectively. Rearranging:

$$V = IR + E$$

and multiplying through by I gives:

$$IV = I^2R + EI \qquad (3.10)$$

Now, the left-hand side of this equation is the power supplied by the source to the motor. The first term on the right-hand side, I^2R, is the rate of energy lost in the coil due to resistance heating. But the total electrical power supplied, IV, goes partly into heat and partly into useful mechanical power. Hence, ignoring other minor power losses, the final term, EI, must represent the mechanical power.

Exercise 3.15

The resistance of the armature in a d.c. motor is usually kept small, around 1 Ω or less. Explain why this is a good idea.

Exercise 3.16

Look back to Table 3.1 in Section 4.4, showing the technical specifications of my small d.c. motor. Note that the no-load current is a lot smaller than the 'maximum efficiency current' for this motor. The resistance of the armature coil can vary a little under different operating conditions, but that's not the cause of the big difference in current here. What is?

5.3 What about a.c. motors?

So far I have used d.c. motors to help you appreciate how electrical energy can be harnessed to generate forces on a small and large scale. However, you have probably noticed that most motorized electrical appliances, particularly the larger ones, don't run off the direct current from chemical cells. They use alternating current that is supplied through household mains sockets instead. Are these a.c. motors the same? Well, in some cases they are, but generally they are not. You might expect that a d.c. motor wouldn't work with an a.c. supply, at least not without some physical modification. The basic problem is that the commutator, which swaps round the current direction every half turn in each set of armature coils, just isn't needed for an a.c. supply. In fact *universal* motors, which run on both a.c. and d.c., get round this by switching the magnetic field as well. If the field coil is an electromagnet then it can be powered using a.c. that reverses the magnetic poles at the same time that the current in the armature is reversed. This is achieved by simply connecting up the armature and field coils in series and powering them from the same a.c. source. However, the efficiency of small a.c. motors of this type is low because the varying magnetic flux leads to greater eddy current effects. Mitigating this by incorporating high-quality laminated cores means that universal motors tend to be expensive. On the other hand, compared to other a.c. motors they can be easily made to rotate at high speed, over 20 000 rpm, and are very compact for a given power rating, making them ideal for use in drills, food mixers and fans.

Far more widespread, particularly for industrial use, is the a.c. *induction* motor. It works on the principle that a moving, or changing, magnetic field can set a conductor in motion. In fact it's like the a.c. generator of Figure 3.37 in reverse, with an alternating current fed to the stator in order to make the rotor turn, as shown in Figure 3.38. (Have you noticed how practically everything you have found out about electromagnetism also works 'in reverse'?)

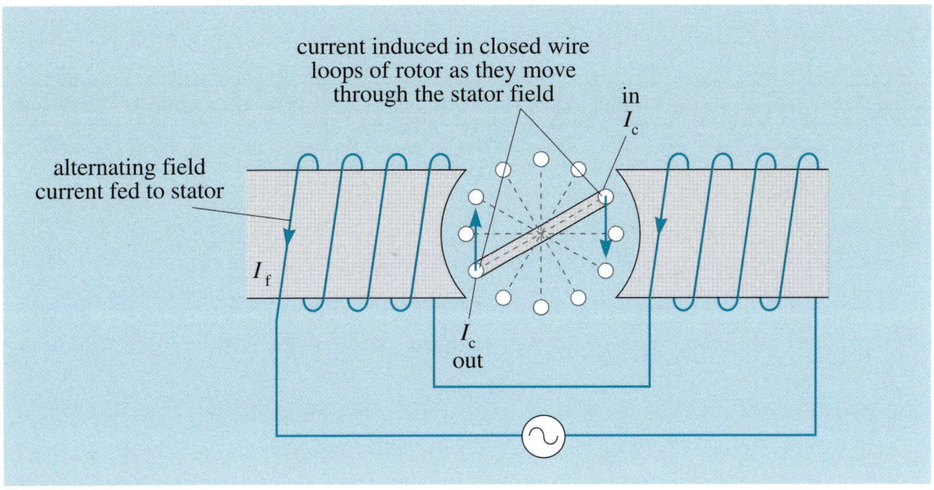

Figure 3.38 An a.c. induction motor. Motion is driven by the interaction of induced coil current I_c and the a.c. stator field produced by the stator current I_f

The rotor of the induction motor consists of 'shorted' wire loops, without external connections. The rotor turns because it experiences a torque due to circulating currents induced in the loops by the changing magnetic field of the stator. These induced currents are just like the eddy currents that were a nuisance in d.c. or universal motors. Here, because the current flow is constrained by the loop, the associated energy is put to good use in turning the rotor, rather than being wasted as heat.

The big advantage of the a.c. induction motor is that we don't need to devise some artificial means of switching the current direction every half turn because the alternating current does that for us – we have a motor that doesn't need brushes! Note that the rate of switching, and hence the rate of rotation of the rotor, depend on the *frequency* of the electricity supply, i.e. how often it changes direction. This is usually fixed (at 50 cycles per second in the UK), and so induction motors are essentially *constant speed* devices. Electronic speed control that provides a variable-frequency power supply (and hence variable speed of operation) can be incorporated, but at extra expense.

In practice, the rotor used in induction motors often has a special design that allows the induced eddy currents to flow in large circuits as shown in

Figure 3.39. It is made up of a number of copper rods in an iron cylinder or 'squirrel cage' – so-called because it resembles a cage for exercising squirrels!

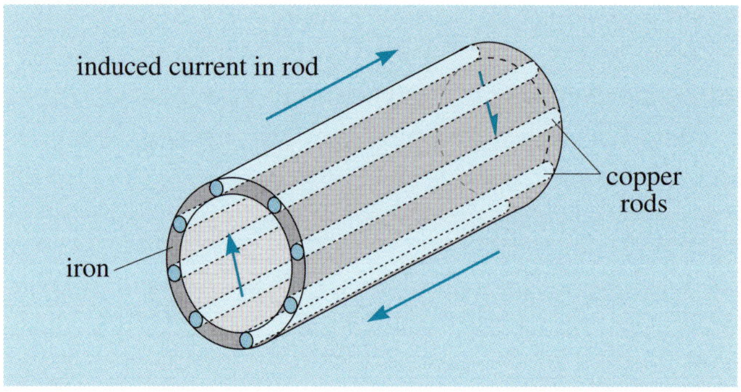

Figure 3.39 Schematic diagram of a squirrel-cage rotor for an a.c. induction motor. Note that the conducting rods are electrically connected with end rings, not shown here

The striking thing about the squirrel-cage rotor is its remarkable simplicity. There are no complicated coil windings or connectors, just a series of conducting metal bars arranged in a cylinder. The design is both efficient and cheap to construct and maintain.

Many large a.c. induction motors are made more powerful by using three pairs of stationary electromagnets in the stator, each powered separately using *three-phase* alternating current. This type of a.c. supply, used widely in industry but not available in your home, provides three separate a.c. waveforms each shifted by one-third of a cycle (Figure 3.40). Using these three currents simultaneously allows constant high power to be generated.

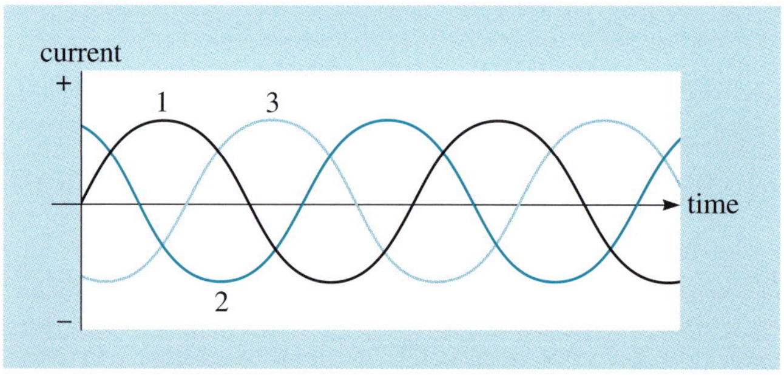

Figure 3.40 Three-phase alternating current

As I have pointed out squirrel-cage induction motors, unlike d.c. motors, have no commutator, and hence no brushes to wear, thus providing better efficiency and reducing maintenance. Moreover, because there are no moving electrical contacts, sparking does not occur and a significant fire hazard is removed. These are some of the reasons why induction motors are preferred for large-scale industrial tasks. A typical industrial design, capable of supplying hundreds of kilowatts of power, is shown in Figure 3.41.

Note again how the heat generated by such a powerful machine must be dissipated by encouraging the flow of air over internal components using a cooling fan. Smaller, single-phase a.c. motors are also frequently used to operate machines where an output of less than one horsepower (746 W) is required. They are, however, less efficient and more expensive to operate than three-phase machines.

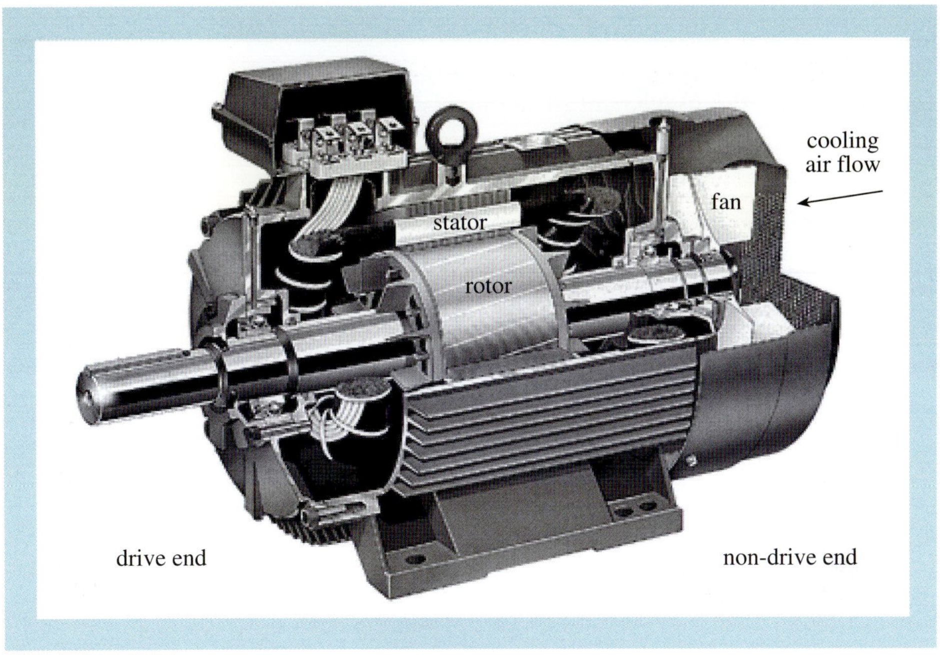

Figure 3.41 A squirrel-cage induction motor

During the early part of the twentieth century d.c. motors were ubiquitous, but they have gradually been replaced, mainly by a.c. induction motors, basically because of efficiency and maintenance benefits associated with the elimination of the commutator. So why do we still have d.c. motors at all?

Exercise 3.17

Considering all that you have learnt about motor design, what are the *advantages* that the d.c. motor has over the a.c. induction motor that have saved it from elimination?

In fact, the d.c. motor has regained some of the ground it has lost to the a.c. induction motor over the years by presenting itself in *brushless* form, i.e. without a commutator. This is achieved by making the armature coil stationary and moving it to the outside of the permanent magnet, which is in turn moved to form the central shaft. The current direction still needs to be swapped over as the shaft rotates, but this is achieved by electronically switching the current flow using a *motor controller*. Most computers use brushless d.c. motors in their rotating disk drives since they have now become smaller and less expensive than their a.c. counterparts.

But of course a.c. motor design hasn't remained static either. For example, the use of *linear* induction motors has increased greatly in recent years. This device is basically a squirrel-cage induction motor opened out flat. In other words, the magnetic field, instead of rotating, now sweeps across the flat motor face. Rather than producing rotary torque from a cylindrical machine, linear force is created from a flat one. Applications include sliding doors, fairground rollercoasters and futuristic 'MagLev' (magnetically levitating) railways.

Finally, be aware that there are many other types of motor design, both a.c. and d.c., that I haven't even mentioned in these pages. Furthermore, some motors use neither continuous a.c. nor d.c. to make them work. *Stepper motors*, for example, translate discrete electrical *pulses* into rapid bursts of precise mechanical movement. These motors are useful for accurately aligning imaging devices such as cameras and telescopes, for example. They are also ideal for coping with the complex acceleration and adjustments associated with the movement of the heads in computer disk drives and inkjet printers.

5.4 Summary of Section 5

- Larger motors are more complex, more powerful and more efficient than smaller ones.
- Generators work on the principle of electromagnetic induction: current will flow in a wire loop when it is moved in a magnetic field.
- Industrial a.c. generators use rotating electromagnets, driven by turbines, to induce large currents in external static coils.
- The a.c. induction motor is like an a.c. generator in reverse and uses an alternating magnetic field to set a conductor in motion; it requires no commutators or brushes.

6 CONCLUDING REMARKS

I began this part of the Dynamics block by talking about Stone Age axes, and finished with complex modern electric machines. Hopefully you have been able to see through the later complexity to keep sight of the underlying force-generating principles that drive the motors and generators I have described. In effect, an electromagnetic force provides leverage causing a shaft to be turned. I wonder what Hero of Alexandria would have made of all this?

7 LEARNING OUTCOMES

After studying Part 3 of Block 3, you should be able to do the following.

3.1 Describe and interpret the 'motor effect' both qualitatively and quantitatively. (SAQs 3.1, 3.2 and 3.3)

3.2 Quantify the efficiency of a motor by comparing electrical power input with mechanical power output, and distinguish between high efficiency and high power output. (SAQs 3.4 and 3.11)

3.3 Calculate the mechanical torque and power associated with a simple d.c. motor. (SAQs 3.4 and 3.8)

3.4 Interpret the linear torque–speed equation for a d.c. motor in graphical form, and also solve it numerically for either torque or speed. (SAQs 3.5, 3.6 and 3.8)

3.5 Use the maximum power equation for a d.c. permanent magnet motor and recognize it as the peak of the power–speed curve. (SAQs 3.7 and 3.9)

3.6 Select appropriate gear ratios in order to magnify motor torque. (SAQs 3.8 and 3.10)

3.7 Recognize and use the quadratic power–speed equation for a d.c. permanent magnet motor, and solve it graphically for ω. (SAQs 3.9 and 3.10)

3.8 Understand and interpret electric motor performance data, and use it to optimize the design of a motor-driven mechanism. (SAQs 3.4, 3.5, 3.7, 3.8 and 3.10)

ANSWERS TO EXERCISES

Exercise 3.1

The axe consists of a wedge, or two inclined planes (the flint), attached to a lever (the handle).

Exercise 3.2

(a) Wheel and axle, screw and (bevel) gears. There is also a lever effect when holding the drill steady.

(b) Lever, wheel and axle, screw and (rack and pinion) gears.

(c) Lever and screw.

Exercise 3.3

(a) **d**

(b) **g**

(c) **c** and **f**

(d) **b** and **e**

(e) **a** and **d**

Exercise 3.4

(a) The wire will move out from the magnet towards the bottom right-hand corner of the figure.

(b) The obvious ways you might have guessed are to use a stronger magnet or a larger current, but having a longer length of wire within the magnetic field also increases the total force.

Exercise3.5

The units of B are N A^{-1} m^{-1}.

Exercise 3.6

If the current were switched off the force on the wire would be removed, because it no longer has a magnetic field associated with it. However, it would still keep moving – recall Newton's first law. If the current were reversed, the wire would decelerate and then start to move back the opposite way because the imbalance of forces would be reversed. Convince yourself this is true by imagining the current direction is reversed in Figures 3.12 and 3.14.

Exercise 3.7

To ensure that the long direction of the laminations doesn't cut across the lines of magnetic flux. Induced currents are thereby restricted to flow in the short thickness of each plate.

Exercise 3.8

Recall that:

$$\text{torque} = \text{turning force} \times \text{distance from axis of rotation}$$

hence the units must be of the form 'force × length' not 'mass × length' (g cm) as given in the table. Correct SI units are N m (or N cm etc.).

If $g = 9.8$ m s^{-2} then the force or weight due to a 1 kg mass on Earth is 9.8 N, so:

$$1 \text{ g cm} = 0.001 \text{ kg cm} = 0.0098 \text{ N cm} = 0.000\,098 \text{ N m } (\approx 10^{-4} \text{ N m})$$

or

$$1 \text{ N m} = 10\,200 \text{ g cm}$$

Exercise 3.9

The largest torque that can be generated by the motor is Γ_s. To find the corresponding mass:

$$\Gamma_s = mg \times r$$

$$4.67 \times 10^{-3} = m \times 9.8 \times 0.01 \text{ N m}$$

$$m = \frac{4.67 \times 10^{-3}}{9.8 \times 0.01} = 0.0476 \text{ kg}$$

Exercise 3.10

(a)

$$\begin{aligned} \text{power} &= \text{rate of work} \\ &= \frac{\text{force} \times \text{distance}}{\text{time}} \\ &= \frac{(0.25 \times 9.8) \times 1}{5} = 0.490 \text{ W} \end{aligned}$$

(b) Yes, the motor can do work at 1.76 W, from a 3.0 V supply.

Exercise 3.11

(a) A, B and v. These all appear on the bottom of the term on the right-hand side of Equation (3.9). Hence making them larger causes P_{therm}/P_{mech} to be smaller. This means greater efficiency.

(b) B and v. From Equation (3.7) you should see that increasing either of these would increase the mechanical power output.

Exercise 3.12

Some of the more notable changes in the larger motor design are:

- there are many more coils in the armature;
- similarly, the commutator is divided into numerous sections;
- the permanent ‘field magnet’ has been replaced by a ‘field coil’;
- it is contained in a sturdier enclosure;
- there is a cooling fan attached to the shaft and air vents have been incorporated into the casing;
- the brushes are attached to the outer casing;
- ball bearings are attached to the rotating shaft.

Exercise 3.13

Remember that the copper coil wire must be electrically insulated. The main effect of any temperature rise is a steady degradation of the electrical insulation on the windings.

Exercise 3.14

(a) Power lost is the difference between the input and output power:

$$P_{lost} = P_{in} - P_{out}$$

but the efficiency $\eta = P_{out}/P_{in}$ so $P_{in} = P_{out}/\eta$.

So the power lost can be related to the output power by combining these ideas:

$$P_{lost} = \frac{P_{out}}{\eta} - P_{out}$$

Then for the small motor,

$$P_{lost} = \frac{0.41}{0.44} - 0.41 = 0.52 \text{ W}$$

and for the large motor,

$$P_{lost} = \frac{1500}{0.7} - 1500 = 640 \text{ W}$$

(b) The surface area of a cylinder of radius r and length l is:

$$A = 2\pi r^2 + 2\pi rl$$

Table 3.4 summarizes the properties of the two motors.

Table 3.4 Summary of properties of the two motors

Output power / W	Efficiency	Length / m	Radius / m	Thermal power loss / W	Surface area / m^2	Power loss per unit area / W m^{-2}
0.41	0.44	0.025	0.01	0.52	2.2×10^{-3}	240
1500	0.7	0.40	0.10	640	0.31	2060

Notice that the increase in the rate of thermal energy that needs to be dissipated has not been compensated for by the increased surface area of the larger motor.

Exercise 3.15

Low resistance means that the heat dissipation term, I^2R, in Equation (3.10) is also low. Hence the motor is more efficient.

Exercise 3.16

The lower current occurs when the speed of rotation is greater (no load). The induced e.m.f. is larger at greater speeds and hence provides greater opposition to the current, which is therefore lower.

Exercise 3.17

A d.c. motor can be easily run off the direct current from a chemical cell, thus making it portable. Its simplicity means that it is also cheap and easily manufactured to a small size. Hence d.c. motors are still a necessity for battery-driven applications (including forklift trucks and milk floats, as well as toothbrushes!). In many applications d.c. motors also offer a cheap and desirable flexibility in terms of easily controlled torque and speed.

ANSWERS TO SELF-ASSESSMENT QUESTIONS

SAQ 3.1

$$F = BIl\sin\theta = 0.2\times2.0\times0.05\times0.707 = 0.014\ \text{N}$$

SAQ 3.2

(a)

$$B = 0.3\ \text{T},\quad I = 4.0\ \text{A},\quad l = 0.1,\quad \theta = 90°\quad \text{so}\quad \sin\theta = 1$$

$$F = BIl\sin\theta = 0.3\times4.0\times0.1\times1 = 0.12\ \text{N}$$

(b) If the force, F, is halved then, from Equation (3.2) the quantity $BIl\sin\theta$ must also be halved. But B, I and l remain constant so it is actually the magnitude of $\sin\theta$ that is halved:

$$\sin\theta = \tfrac{1}{2},\quad \theta = 30°$$

However, this is the angle between the wire and the field that halves the force (see Figure 3.16). The wire must be *rotated* through 60°.

SAQ 3.3

The correct word order is:

positive, downwards, magnetic flux, current flow, force, anticlockwise, inertia, negative.

SAQ 3.4

Nominal voltage	3.0 V
Electrical power input / W	$IV = 0.96\times3.0$ $= 2.88$
Angular speed / rad s^{-1}	$\omega = \dfrac{11300}{60}\times2\pi$ $= 1180$
Torque / N m	$\Gamma = 10.4\times9.8\times10^{-5}$ $= 1.02\times10^{-3}$
Mechanical power output / W	$\Gamma\omega = 1.02\times10^{-3}\times1180$ $= 1.20$
Maximum efficiency	$\dfrac{1.20}{2.88} = 0.42$

SAQ 3.5

From the completed Table 3.2, for maximum efficiency at 3.0 V:

$$\omega = 1180\ \text{rad s}^{-1}$$

when $\Gamma = 1.02\times10^{-3}$ N m

and from Table 3.1, at no load and 3.0 V:

$$\omega = \omega_n = \frac{14400\times2\pi}{60} = 1510\ \text{rad s}^{-1}$$

when $\Gamma = 0$

Hence the gradient

$$\frac{d\Gamma}{d\omega} = \frac{0 - 1.02\times10^{-3}}{1510 - 1180} = -3.09\times10^{-6} \quad \left[= -\frac{\Gamma_s}{\omega_n} \right] \quad \text{N m s rad}^{-1}$$

and stall torque

$$\Gamma_s = 3.09\times10^{-6}\omega_n = 3.09\times10^{-6}\times1510 = 4.67\times10^{-3}\text{ N m}$$

SAQ 3.6

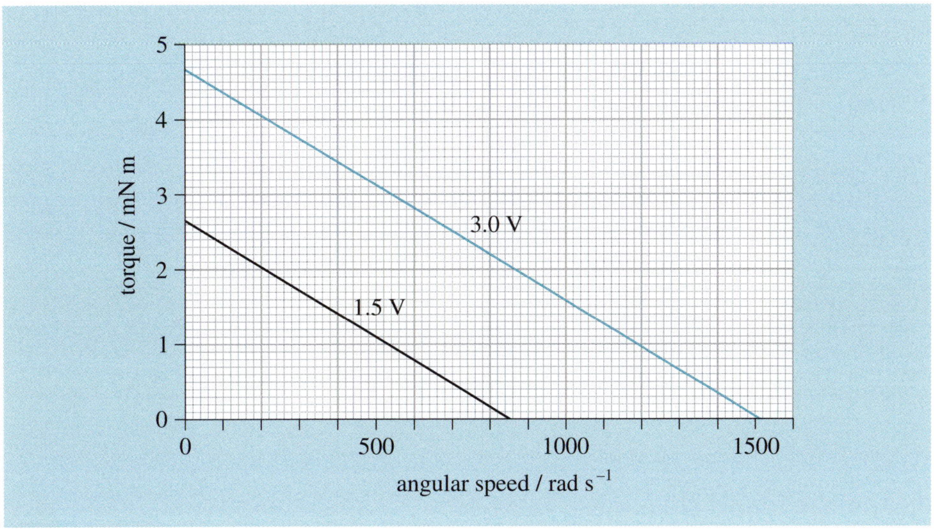

Figure 3.42

SAQ 3.7

Using Equation (3.6),

$$P_{max} = \tfrac{1}{4}\omega_n\Gamma_s = \tfrac{1}{4}\times1510\times4.67\times10^{-3} = 1.76\text{ W}$$

SAQ 3.8

Gear ratio	4:1	16:1	64:1	256:1
Shaft torque Γ_{motor} / N m	6.13×10^{-3}	1.53×10^{-3}	3.83×10^{-4}	9.57×10^{-5}
Shaft speed ω_{motor} / rad s^{-1}	stall	1020	1390	1480
Power P / W	–	1.56	0.532	0.142
Time for lifting t / s	–	1.57	4.61	17.3

The 64:1 gear ratio is most suitable since it allows the book to be lifted in about 5 s. (Note that using the 4:1 gear train would generate a torque larger than the motor stall torque).

SAQ 3.9

(a) The values you read from the graph should be about 1400 rad s^{-1} and 100 rad s^{-1}. Note that using the graph isn't as accurate as solving the equation numerically.

(b) $P_{max} \sim 1.76$ W

SAQ 3.10

(a) The motor is not experiencing a load while the train wheels spin at 120 rpm. This is equivalent to $120/60 = 2$ revs per second or $2 \times 2\pi = 12.6$ rad s^{-1}. Comparing this to ω_n we have:

$$\frac{\omega_n}{12.6} = \frac{860}{12.6} \approx 68$$

i.e. the train wheels spin about 68 times slower than the motor shaft. Hence the 64:1 gear ratio has probably been used.

(b) The speed of the train wheels is slightly slower than expected because of efficiency losses in the gear train.

(c) Putting $\omega = (3/4)\omega_n$ into the power–speed Equation (3.4) we get:

$$P = \Gamma_s\omega - \frac{\Gamma_s}{\omega_n}\omega^2 = \Gamma_s \tfrac{3}{4}\omega_n - \frac{\Gamma_s}{\omega_n}\left(\tfrac{3}{4}\omega_n\right)^2 = \tfrac{3}{4}\Gamma_s\omega_n - \tfrac{9}{16}\Gamma_s\omega_n$$

$$= \left(\tfrac{3}{4} - \tfrac{9}{16}\right) \Gamma_s\omega_n = \tfrac{3}{16} \times 2.6 \times 10^{-3} \times 860 \text{ W}$$

$$= 0.42 \text{ W}$$

SAQ 3.11

(a) The mechanical output power per conductor in the coil is $P_{mech} = BIlv$, in which l is the effective length of the coil (rotor), so doubling length, doubles power.

(b)

$$\frac{P_{therm}}{P_{mech}} = \frac{\rho I}{BvA}$$

must be kept small for high efficiency; in this expression A is the cross-sectional area of the wire in the rotor coil; large A, is favoured for high efficiency.

ACKNOWLEDGEMENTS

Grateful acknowledgement is made to the following sources for permission to reproduce material within Part 3 of this block.

Figures

Figure 3.4: George Bernard/Science Photo Library; *Figure 3.11:* source: http://www.chem.ch.huji.ac.il; *Figures 3.12, 3.14, 3.19, 3.20, 3.22 and 3.35:* adapted from Johnson, K. (1991) *Physics for You,* Nelson Thornes Ltd; *Figure 3.16(a):* Breithaupt, J. (1991) *Understanding Physics for A Level,* Nelson Thornes Ltd; *Figure 3.18(a):* source: http://www.chem.ch.huji.ac.il; *Figure 3.18(b):* Nelkon, M. (1971) *Fundamentals of Physics,* Chatto and Windus; *Figure 3.32:* GE Support Services; *Figure 3.34:* Reliance Electric; *Figure 3.41:* Courtesy of ABB Asea Brown Boveri Ltd.

Every effort has been made to contact copyright owners. If any have been inadvertently overlooked, the publishers will be pleased to make the necessary arrangements at the first opportunity.

Course team acknowledgements

Part 3 of this Block was prepared for the course team by Dr Martin Rist. Several of the illustrations have been developed from S207 *The Physical World*, Book 5 *Static Fields and Potentials* (ed. Joy Manners).

Trademarks

Mathcad® is a registered trademark of MathSoft Engineering & Education Inc., USA.

Part 4
Civil aviation – a transport case study

CONTENTS

1 INTRODUCTION

Flight is an example of where the scientific appreciation of how something works has led to the opening up of a whole technology. Air travel revolutionized life in the twentieth century and continues to set the pace of commerce in the twenty-first. Is 'being there' so important these days? Or has the advance of information technology (IT) diminished the hunger for air transport? Far from it! In fact, IT has been brought to bear on many of the problems of 'getting there'.

The study of how to obtain enough lift from a wing to provide a useful aircraft takes us into the selection of materials and geometries for lightweight structures, into choices of power sources for pushing the wing through the air. Sorting out those issues was enough for early aviators like Blériot (Figure 4.1), but designing a commercial airliner has more to it than making big machines fly. Fuel economy, payload and range are obvious concerns, but maintenance schedules, turn-around time on the ground and flexibility of service are vital for a commercial operation.

Figure 4.1 Blériot's channel-hopping machine. Weighing 300 kg, the 1909 monoplane (Blériot model XI) was made on a frame of ash, bamboo and steel. It had a wing span of 7.8 m and a wing area of 14 m^2; its maximum speed was 58 km h^{-1} (about 36 miles per hour)

The A340 airbus, for example, has made big strides in all these areas, not least by developing a better aerofoil section for the wings, but also through judiciously deploying new materials and by designing modular cabin systems that allow rapid changes of function. Cynics, however, might say that credit deals are the best seller of all – and they may be right!

In this concluding part of Block 3, I want to explore some of the challenges faced by the commercial aviation industry. At the heart of the story is the question of how to meet the steadily increasing demand for air miles – arising from both more destinations and more passengers (Figure 4.2). The routine solution would be to build more aircraft and construct more airports, but the industry has recognized a need for innovative solutions that escape the constraints of a system that appears to be approaching saturation.

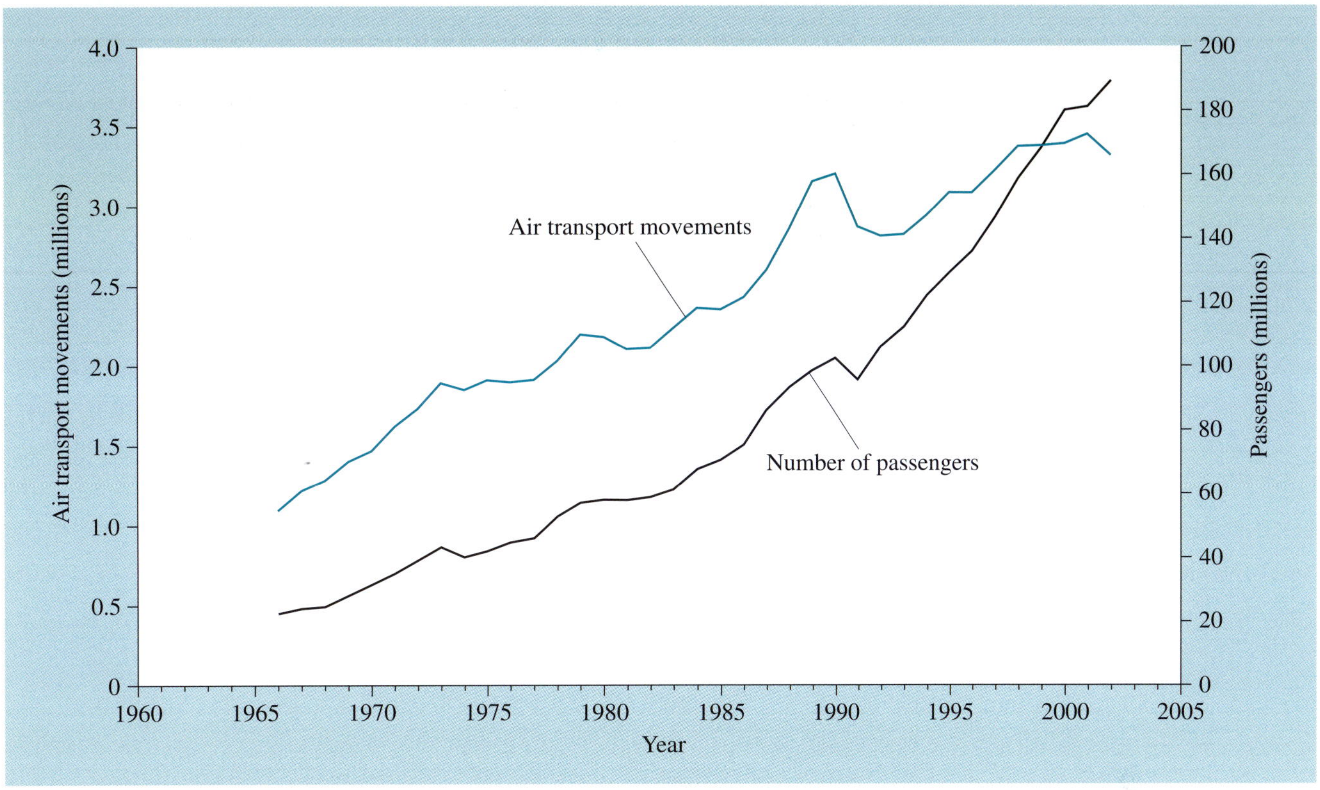

Figure 4.2 Demand for civil aviation in the UK – air transport movements and number of passengers. *Source:* Civil Aviation Authority, UK

1.1 Aim

The aim of this final part of Block 3 is to explore some of the problems faced by the commercial aviation industry in meeting the steadily increasing demand for passenger miles.

It is necessary to understand some of the technical aspects of flight to appreciate the scope for innovations, so we will be exploring some of the fundamentals of flight. In a nutshell they are these. Flying involves lifting a body into the air against the pull of gravity, while moving through the air against the drag force that accompanies such motion. Simple – it sounds like the stuff of dynamics? We shall see.

2 SETTING THE SCENE

Civil aviation is a mature business with global dimensions. It has evolved to meet one of the major needs of our trading society – the need to get around between the spaces we inhabit. It is important to begin by considering the environmental and commercial contexts in which our engineering is to be applied.

2.1 The environmental context

The first passenger-carrying flight is usually reckoned to have taken place in Paris on 19 September 1783. However noteworthy an achievement this was, it can scarcely be described as the event that ushered in the era of the global civil aviation industry that we know today – the flying machine was a hot-air balloon and the passengers were a cockerel, a duck and a sheep. Over a century later, in 1910, the first regular, commercial, passenger-carrying flights started – an intercity airship service in Germany. International routes flown by aeroplanes had to wait until after the First World War, benefiting from the developments in technology that were accelerated by that war. They began in 1919, first between Paris and Brussels, and this was followed a month or so later by a daily service between London and Paris. The decades that followed have seen civil aviation develop from a mode of transport for the few who could afford it (and were either brave enough or foolhardy enough to risk it, depending on your point of view) to the mass transport system encircling the globe that we have today.

This vast expansion has also seen ticket prices plummet, so that flying has become affordable to the masses, particularly in the relatively affluent Western world. Destinations that were previously unattainable have become feasible, whether for holidays or business trips. It is now cheaper to fly to many places within Europe than to travel by train or road. But we ought to pause and ponder whether there has been a price to pay for all these benefits and all this expansion.

Example

Take a few moments to think about some of the disbenefits that might be associated with civil aviation. Jot down your thoughts in terms of the effects on:

individual passengers

communities near airports

nations

the world.

Answer

Here are my thoughts:

- **individual passengers:** air sickness, jet lag, deep vein thrombosis, fear of flying (including fear of acts of terrorism such as hijacking), air rage, general discomfort and boredom, keeping children amused in confined spaces, baggage ending up at a different destination;

- **communities near airports:** pollution (including noise), anxiety about crashes (take-off and landing are the riskiest stages of flying), surface traffic congestion, likelihood of local airport expansion, possible decrease of property values;
- **nations:** cost of maintaining 'flag-carrier' airlines for national prestige, risk of import of infectious diseases, cost of infrastructure (e.g. airports, air traffic control, surface transport links to airports, policing and customs), problem of how to cope with expansion of existing airports and/or siting of new ones, opening up of new holiday areas risking damage to environment and wildlife, inability to impose a 'green tax' on aviation fuel without international agreement;
- **the world:** aircraft consume large quantities of fossil fuels and are significant contributors to atmospheric pollution especially the so-called 'greenhouse gases', air travel can bring about the rapid spread of infectious diseases.

Your list may well differ from mine, but you should have got the point. It reinforces an important message about engineering: that it is virtually impossible to do any form of engineering that has no consequences, whether direct or indirect, on other people or the environment. Any ethical dimension to engineering (and to individual engineers!) requires that such consequences be thought about and be taken into account, for example by means of an all-embracing approach to cost–benefit analysis.

2.2 The commercial context

Before we investigate the engineering issues there is more to learn about the 'need' that gives rise to the challenge. The business of civil aviation is a truly global enterprise and an audio programme *Up in the Air* from a radio series on Global Business is available on the T207 CD-ROM to set the commercial context. Listen to it as soon as possible and before you get to Section 4.

Civil aviation is a business that turns over vast sums of money. We need to think about the nature of the costs involved to see where the engineer, as opposed to the economist, might be the problem solver.

Exercise 4.1

The costs associated with a national airline include the cost of lease or purchase of aircraft, operational costs and the costs of the marketing infrastructure. Highlight at least three areas where the subjects of engineering materials and engineering dynamics have some relevance.

The scope for engineering lies in the origination of new flying machines and the adaptation of existing apparatus. So where do new airliners come from? The development of modern airliners needs vast capital investment and is centred around the assembler of the airframe – the part to which other bits like wings, engines and tail are ultimately connected. In fact, there are now really only two airframe manufacturers that are expected to deliver major passenger airliners in the foreseeable future (Airbus and Boeing). How do they compete?

Well, given that the number of passengers is steadily increasing and that there are limits to the capacity of airports, air space and ticket price, a contest is emerging for the provision of aircraft that enable higher throughput. Notice the carefully chosen words – the solution need not involve larger aircraft.

The contest extends into ways of paring down operational costs through efficiency gains, some of which may call for new aircraft, new engines or new business infrastructure.

It's a fickle business. The aircraft are subject to the rules of engineering and the laws of science, whereas the behaviour of passengers and shareholders is somewhat less easily determined. Let's get back to safer ground and find out some more about flight.

3 BASIC FLIGHT MECHANICS (IS IT A BIRD?)

Ever since people have been able to think about it, they must have wondered why they could not fly like so many creatures around them. Nothing could stop them trying, though. Not surprisingly, most of the early attempts to build flying machines were based on the principles of bird flight, using flapping wings. As far as we know, these attempts have always been futile, perhaps with the exception of Daedalus and his son Icarus; according to Greek myth, they flew with wings made of quill feathers joined together by wax. Legend has it that their flight ended in tragedy when the wax in Icarus's wings melted since he flew too close to the Sun! (In fact, temperature gradually drops with altitude in the lower layers of the atmosphere.)

Birds flap their wings energetically to be able to fly (up to 80 flaps per second in the case of hummingbirds). Complex movement of the wings provides both a keeping-up function (*aerodynamic lift* or simply *lift*) and a forward motion (*thrust*), as well as enabling manoeuvres.

An energetically more efficient way of flying is used by migrating birds, such as storks and geese, which travel thousands of miles and do not have the energy to flap their wings all the time. They use thermals (rising currents of warm air) to gain height (and hence *potential energy*) and glide, with their wings firmly stretched, to cover horizontal distances (converting potential energy to *kinetic energy*). Of course, they lose height in the process, but that can be recovered in the next thermal. This idea of gliding, as opposed to flapping wings, was taken up by Sir George Cayley, an English engineer who laid the foundations of the modern aeroplane in a series of published technical articles in 1809. He proposed a flying machine that had fixed wings to provide the 'keeping up' function and an engine for propulsion. This separation of the functions of thrust and lift, simple though it sounds, was a revolutionary change in the way people thought about flying machines. He even added a tail to his design to help control stability. Unfortunately, he did not have the technology at the time to build such a self-propelling flying machine, but in 1853 he was successful in building the first human-carrying glider in history.

We had to wait another fifty years for the American Wright brothers to build the first successful manned, power-driven, heavier-than-air flying machine. Their plane, powered by two propellers, chain-driven by a petrol engine, was initially airborne for 12 seconds, but later flights on the same day lasted a bit longer.

Only 80 years after the Wright brothers' historic flight, I entered into the waiting lounge for a long-haul flight and saw the aeroplane that I was shortly to board. It was a Boeing 747 jumbo jet and it seemed huge. There were nearly 400 passengers in the lounge – which made me more than a little nervous. How on earth could this beast fly, carrying all these passengers, their luggage and some cargo, plus fuel and provisions (the total mass of the aeroplane with full load can be up to 360 tonnes)? Well, we will now look into the basic flight mechanics of heavier-than-air aeroplanes to find out.

A brief history of civil aviation is given on the T207 CD-ROM in the form of an animated 'timeline'.

We will be looking at the forces involved in flight and how they arise from airflow over surfaces and from engines. We will also consider how to balance the forces and movements to achieve control in flight.

3.1 Forces of flight

It is a good idea to start by describing the forces that an aeroplane is subjected to in order to keep it in steady flight. We know that, to remain airborne, an aeroplane, or a bird for that matter, requires an upward force, i.e. lift. This counterbalances the force of gravity, namely the weight. We also know that the wings can only generate lift when they move through the air and for this we need thrust, which is provided by the engines of the aeroplane, or the flapping of the bird's wings. There is, unfortunately, a fourth element, working against a flying object: air resistance or drag, experienced by any moving object in the atmosphere, as you came across in Part 2 of this block. Thus, the flight of an aeroplane involves four basic forces: *lift*, *weight*, *thrust* and *drag*, as shown in Figure 4.3.

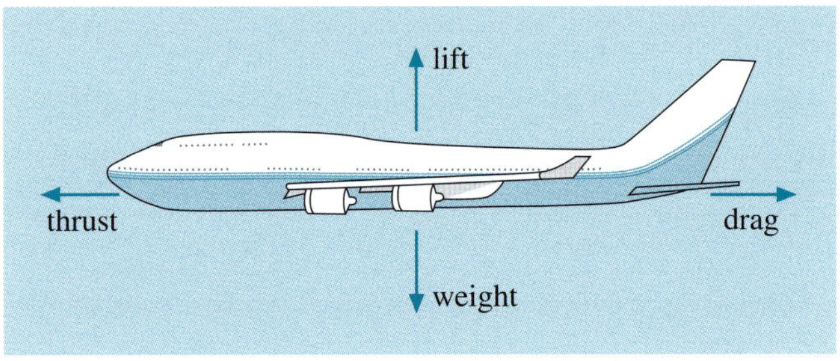

Figure 4.3 Four forces of flight

A simple equilibrium of these four forces gives us the conditions for a steady and level flight. Keeping the aeroplane travelling at a steady height and constant speed requires a lift F_L that exactly matches weight mg and a thrust F_T that exactly matches drag F_D, i.e.:

$$F_L = mg \text{ or } W \tag{4.1}$$

$$F_T = F_D \tag{4.2}$$

Aeroplanes do not always fly steadily, straight and level, though. For example, during take-off the engines must produce sufficient power to exceed the drag, thus allowing acceleration; similarly, the lift must exceed the weight before the aeroplane can become airborne and gain altitude.

Exercise 4.2

Predict the effect of the following flight conditions on an aeroplane.

(a) lift > weight

(b) thrust > drag

(c) weight > lift

(d) drag > thrust

This covers the equilibrium in horizontal and vertical directions. But, as you know from Block 2, there is a second condition for equilibrium. If the forces on a rigid body do not act through a common point, there may be net moments, tending to cause rotation of the body. In level flight, the total moment must be zero to prevent the aeroplane from rotating. However, turning moments are required when it comes to manoeuvring the aeroplane. Stability and control of an aeroplane will be considered in Section 3.7.

Two of the basic forces, weight and thrust, are the physical attributes of the aeroplane. In general, they are known or can be easily determined and controlled. Lift and drag, on the other hand, arise as a consequence of the movement of the aeroplane through the air and are more complicated to determine. The origin of these forces is fundamental to our understanding of flight mechanics, so we will study them in more detail.

3.2 Lift and drag

How does a wing generate enough upward force to support the weight of the aeroplane?

The answer to this lies in the interaction between a wing and the flow of air over it. Once, it was thought that this interaction was simply air particles hitting the bottom surface of the wing and deflecting downwards. The explanation for the upward force on the wing was given using Newton's Third Law: *every action produces equal but opposite reaction*, so this is sometimes referred to as the Newtonian theory of lift. As the particles are deflected downwards by the bottom surface of the wing, they exchange some of their momentum with the wing, pushing it upwards and backwards (Figure 4.4). The resolution of the force arising from the rate of change of momentum of the air into components with respect to the direction of the airflow gives the lift and the drag.

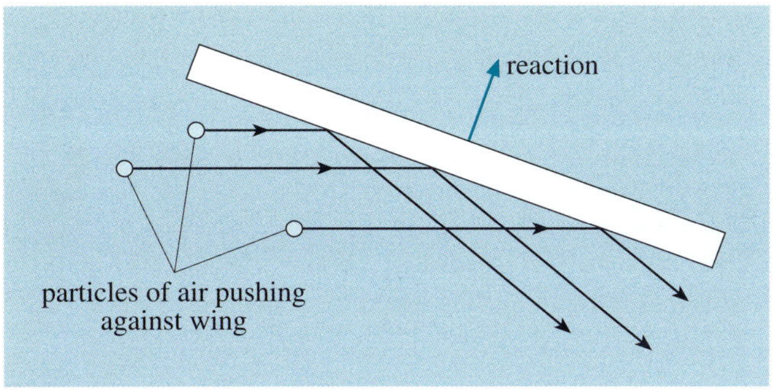

Figure 4.4 Newtonian lift

The mistake in this argument is that it does not take into account that **▼Air is a fluid▲**. Air molecules do not simply behave like bullets from a shotgun. Any fluid that a solid object is moving through will be deflected even before it hits the object, a fact that was observed by Leonhard Euler, the eighteenth-century Swiss mathematician. The reason for this behaviour is that the fluid in the path of a moving object is 'warned' of the arrival of the object by areas of increased and decreased air pressure around and ahead of it – you will have experienced something similar if you've ever tried to swat a fly. These pressure variations are sent in all directions to the surrounding fluid as waves. These pressure waves are effectively the sound waves that are produced by the aeroplane's passage through the air (not the noise of the engines – perhaps you have 'heard' a glider close overhead). So an object moving through a fluid is effectively preceded by waves travelling at the speed of sound; at sea level, sound waves in air travel at 340 m s^{-1}.

This implies that if an object were travelling through the air faster than the speed of sound (supersonic), then the air molecules would not get any warning and would be hit by the object head on, exactly as suggested by the Newtonian theory. Indeed, although the above explanation of upward force by so-called Newtonian lift does not explain the flight of a conventional aeroplane travelling well below the speed of sound, it does give good predictions in certain flight conditions where the speed is very high. These are the conditions that occur around the NASA space shuttles during the early phases of their re-entry into the Earth's atmosphere at altitudes above about 80 km and at speeds above 4000 m s^{-1}.

▼Air is a fluid▲

Fluids have no shape of their own; they can deform easily and adapt to the shape of any container; furthermore, they flow from one container to another. That means liquids and gases are fluids; even finely divided solids, such as grains of sand, can behave as fluids.

The molecules in air are in constant random motion – they move in all directions. They continuously collide with each other and the surfaces of any object they encounter. Since molecules possess mass, momentum and energy, they impart momentum on impact, creating a force normal to the surface. The sum of the forces of all the molecules striking the surface divided by the area of the surface is what we mean by pressure.

An important property of a fluid, one that is of prime importance in the aerodynamics of flying objects, is its density. At sea level, the density of air is approximately 1.2 kg m^{-3}.

Although fluids are molecular, they are considered as continuous substances, as are solids. Fluids must also obey Newton's laws of mechanics:

1. Air that is still will remain still, while air that is moving will resist any change of speed and direction.
2. To alter the state of rest or uniform motion of air, or the direction and speed of air, a force must be applied to the air, and the more sudden the change of speed or the direction and the greater the air mass affected, the greater the force that must be applied.
3. The application of a force to air produces an equal and opposite reaction force on the surface that produces the force.

Clearly, the exact nature of airflow around a wing needs to be known if we are to find an explanation for lift. But the flow patterns produced by relative motion between the air and a wing are often very complicated and a complete physical description requires a deep understanding of aerodynamics, a branch of fluid mechanics. You will see more on fluid mechanics, which covers all aspects of fluids in motion, later in Block 4. Here, I am going to introduce you to some of the basics of aerodynamics which will give you a better understanding of the forces involved in the flight of an aeroplane.

As in most disciplines, the evolution of knowledge in aerodynamics involves learning through observations, developing theories to attempt to explain these observations and testing these theories by experiments. Early pioneers of aviation quickly realized that they did not know much about the mechanics of flight and that it would take more than watching birds to reveal the secrets of it. They therefore built laboratories equipped with instruments that enabled them to test their flying machines, or parts of them, to improve their designs.

Studies in **▼Wind tunnels▲** show that when air flows past a wing, the air pressure underneath the wing is increased, whereas over the top it is reduced. The net effect of this pressure difference is a force on the wing, pushing it both upwards and backwards. Even a flat plate can generate this pressure difference when it is inclined at a small angle relative to the direction of the airflow, producing a net force as shown in Figure 4.5. It is convenient to define forces in aerodynamics relative to the direction of airflow. Thus, *lift* (F_L) and *drag* (F_D) are the components of this resultant force normal and parallel to the direction of the airflow, respectively. When an aeroplane flies through stationary air, the relative airflow is determined by the speed and direction of travel of the aeroplane.

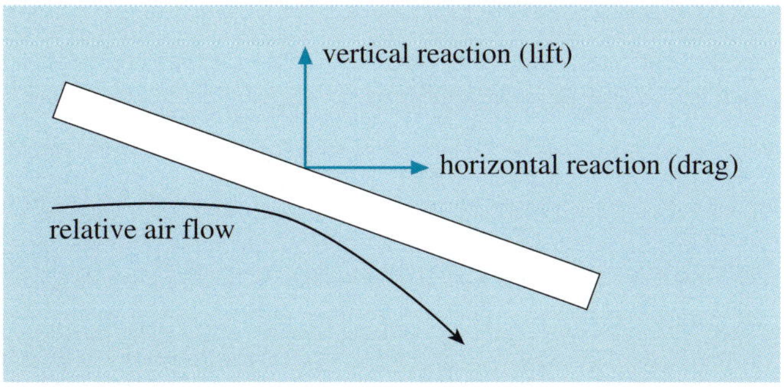

Figure 4.5 Flat plate inclined to airflow, showing resultant force produced

▼Wind tunnels▲

Wind tunnels simulate the conditions encountered by any object moving through the air. A large fan generates a wind that can be controlled in terms of speed, temperature, humidity, etc. For practical reasons, the object studied is held so that it is stationary as air is forced over it. As we are only concerned with the relative motion of the object and the air, the test reproduces the true conditions of atmospheric flight, provided that the dimensions of the object are small relative to the diameter of the wind tunnel. The aerodynamic forces generated by the interaction of the airflow and the object are directly measured as forces on the structure that tethers the object. Likewise, variations of air pressure around the object may be measured through small ports drilled into the object.

To allow visualization of the airflow, traces of smoke are injected into the upstream flow. This helps to establish the relation between the nature of airflow and aerodynamic forces. Wind tunnels are now an essential part of aircraft design, so that no aircraft (including aeroplanes, helicopters, spacecraft and rockets) is ever committed to flight before extensive tests in a wind tunnel (Figure 4.6).

An alternative to making a physical model is to use computers to model fluid flows (computational fluid dynamics). However, wind tunnel tests remain vital for testing the validity of the computer's flow-simulation software. Such software is therefore a valuable complement to wind tunnel tests, rather than a replacement for them.

Figure 4.6 Preparing a wind tunnel test

SAQ 4.1 (Outcomes 4.3 and 4.5)

A large aeroplane having a mass of 200 tonnes is climbing at a constant rate, at an angle of 10°. Calculate the lift force, remembering that it is perpendicular to the relative airflow.

SAQ 4.2 (Outcomes 4.4 and 4.5)

The lift calculated in SAQ 4.1 turns out to be less than the weight of the aeroplane yet it still climbs. Explain why (a simple free-body diagram may help).

The amount of aerodynamic force generated (both lift and drag) increases with the speed of the airflow and the inclination angle to the airflow direction (this angle is called the *angle of attack* and it is an important factor in the aerodynamics of the wings). The flat plate turns out to be an inefficient way of producing lift because it creates a significant amount of drag. The ratio of lift to drag (F_L/F_D) determines the efficiency of a wing in producing the required lift. A low ratio of lift to drag means that more power is needed from the engines to produce enough thrust to overcome the drag associated with the required lift.

The search for more efficient wing sections, i.e. wings with high F_L/F_D, led to the evolution of a thicker section with a cambered top surface, known as an ▼**Aerofoil**▲. A typical aerofoil is shown in Figure 4.7, though, strictly speaking, any lift-producing section may be called an aerofoil, the section of a flat plate being the simplest one.

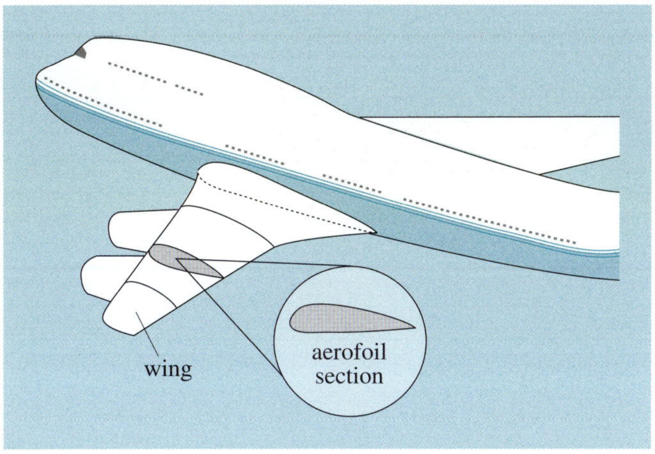

Figure 4.7 Typical aerofoil and its cross-section

▼Aerofoil▲

As well as producing less drag, a thick and cambered aerofoil provides more structural strength and stiffness than a thin flat plate. Although most aerofoils can produce lift even at zero angle of attack, they are often designed to produce the maximum lift at small angles of attack. However, the definition of angle of attack for an aerofoil is not as straightforward as it is for a flat plate.

Figure 4.8 gives some important terminology necessary to describe the geometry of an aerofoil. The front of the wing is called the *leading edge* and the back the *trailing edge*. The straight line drawn from the leading to trailing edges of the aerofoil is called the *chord line* and its length is the *chord*, c. The *camber line* is obtained by plotting all the mean points between top and bottom surfaces from leading edge to the trailing edge, perpendicular to the chord line. The maximum deviation of camber line from the chord line gives a measure of the amount of camber, e. Finally, the angle of attack is defined as the angle between the direction of airflow and the chord line of the aerofoil.

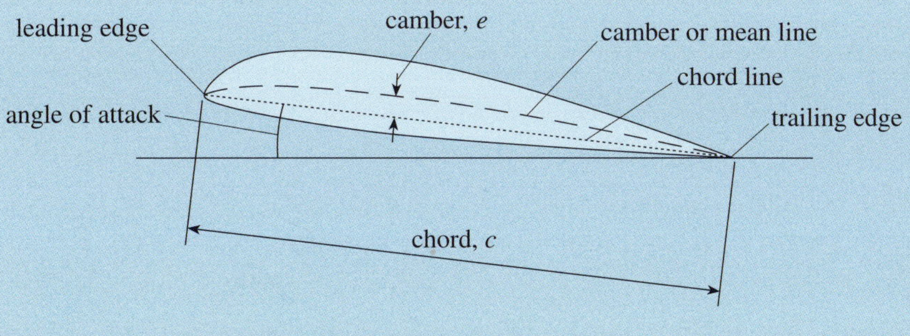

Figure 4.8 Definitions of aerofoil geometry

3.3 The airflow around an aerofoil section

So far, you have learned that the lift is a result of pressure difference caused by the relative movement between air and the wing. Now it is time to see how the airflow around an aerofoil causes this pressure difference. Figure 4.9 shows the *streamline* pattern around a section of flat plate and a typical aerofoil, obtained by the flow visualization methods mentioned above.

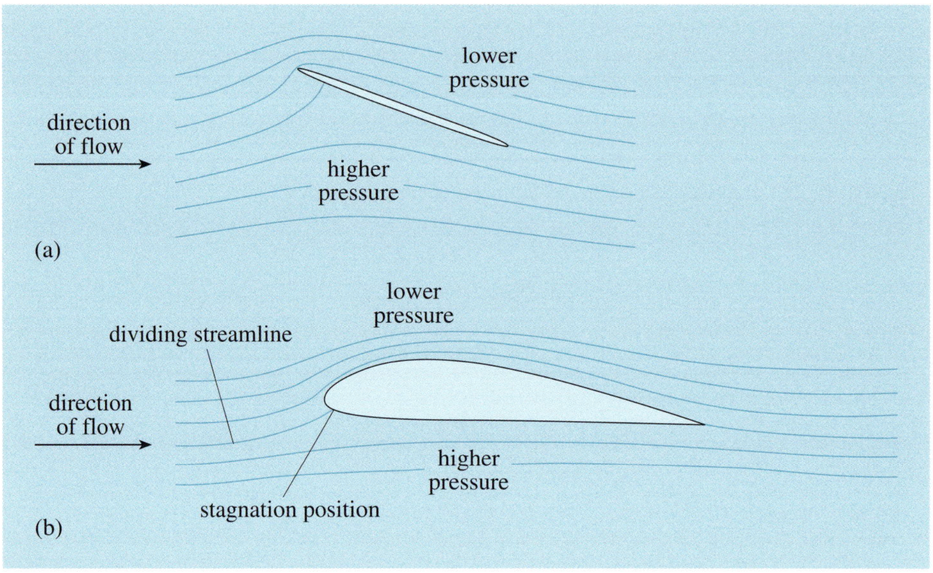

Figure 4.9 Streamlines around (a) a section of a flat plate and (b) an aerofoil

A streamline is an imaginary line drawn to show the path traced by a particle of fluid (provided that the flow is steady, i.e. it does not change with time, unlike in the case of turbulent flow). The direction of motion of fluid particles is always in a direction tangential to the streamlines and there is no flow across them. That is why smoke particles in an air flow can be used to visualize the streamlines. Furthermore, for an ▼**Ideal fluid**▲ the closeness of streamlines is an indication of flow speed. This last claim does not seem obvious; let us explore it with the following example.

Example

Figure 4.10 shows a steady streamline pattern of a continuous flow of fluid flowing through a pipe which is uniform in diameter at both ends but has a constriction in the middle (a so-called Venturi tube). Assume that the fluid is ideal and there is no leak in the pipe. Show that the speed of the fluid at the constriction is faster than the speed at the entrance to the Venturi. (Hint: the flow of mass of a fluid of density ρ moving at speed v crossing any perpendicular plane A is ρvA, with units of kg s^{-1}.)

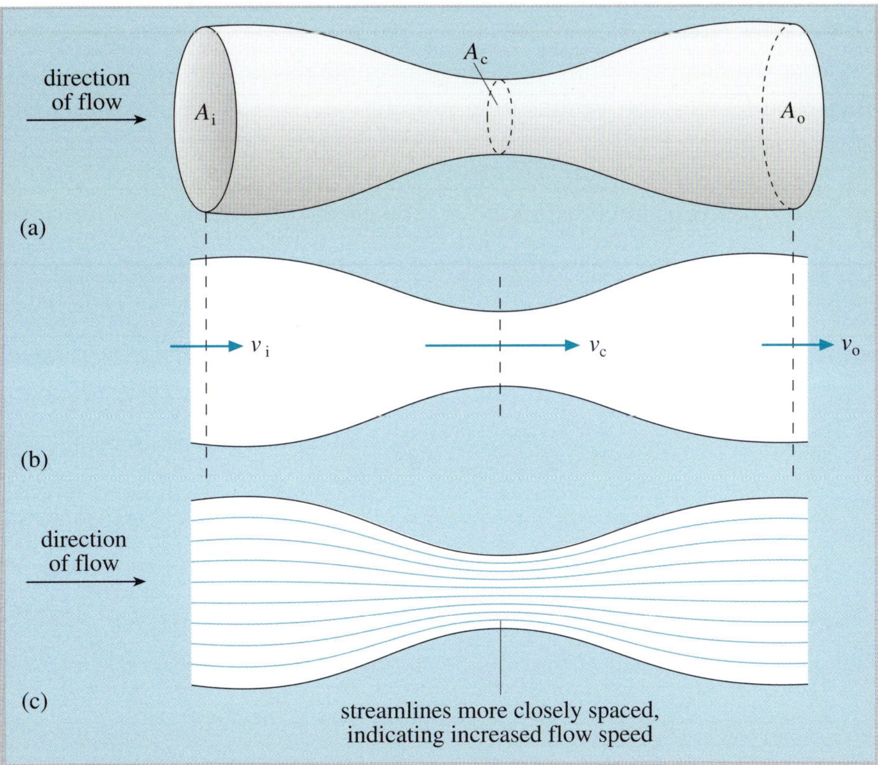

Figure 4.10 Venturi tube – a section where flow is gently constricted

Answer

Since fluid mass passing through the inlet must be equal to fluid mass flowing out from the outlet per unit time, the mass flow rate must also be the same at any section in the pipe. We define mass flow rate $\dot{m}$ (said em-dot) as:

$$\dot{m} = \text{density} \times \text{area} \times \text{speed} = \rho A v$$

If we consider the mass flow rate at the inlet, $\dot{m}_i$, and at the constriction, $\dot{m}_c$, then we can state that:

$$\dot{m}_i = \dot{m}_c$$

or:

$$\rho_i A_i v_i = \rho_c A_c v_c$$

Since the fluid is incompressible (remember, it is ideal), then its density remains constant throughout, i.e. $\rho_i = \rho_c$. The equation then reduces to:

$$A_i v_i = A_c v_c$$

By rearranging the equation, we obtain:

$$v_c = v_i \frac{A_i}{A_c}$$

Since A_i is greater than A_c, it can be concluded that the velocity of the fluid at the constriction is faster than that at the inlet by a factor of A_i / A_c. The general conclusion is that the speed of the flow increases with decreasing cross-sectional area, and vice versa; in other words, the product of cross-sectional area and flow speed (Av) remains constant along a tube (this is sometimes referred to as the continuity principle).

Exercise 4.3

Repeat the analysis in the preceding example to show that the speed of the fluid leaving the Venturi tube in Figure 4.10 is slower than it is in the constriction.

The streamline pattern in the Venturi tube in Figure 4.10 indicates that the streamlines in the constriction crowd closer where the fluid flow is higher. Relatively speaking, widely spaced streamlines indicate regions of lower speed flow, and closely spaced streamlines indicate regions of higher speed flow.

▼Ideal fluid▲

The behaviour of fluids in motion is incredibly complicated and the correct description of this behaviour is essentially its own special branch of mathematics. In order to aid understanding of such physical phenomena, scientists often employ certain assumptions to simplify the basic ideas. Once the underlying physical principles are established, a more precise definition can be sought by systematically introducing real conditions.

How realistically can we simplify fluid flow? Part of the complexity of fluid flow is due to the properties of a fluid; two of the most important ones are its *compressibility* and *viscosity*. We need to know a little more about these properties.

All fluids are compressible to some extent; i.e. their density increases under increasing pressure. Compared with gases, liquids are relatively *incompressible*. In fact, even gases may be treated as incompressible provided the flow speeds involved are not great. For flow over an aerofoil below about 150 m s^{-1}, air may be treated as incompressible (i.e. there is virtually no change in density throughout the flow). However, inside engines, fuel–air mixtures are compressed and an incompressible fluid model is inappropriate.

Viscosity is the resistance of fluid to flow. It is a kind of internal frictional effect exhibited to some extent by all fluids and analogous to the friction between solids in motion. A sluggish liquid, such as extra virgin olive oil or golden syrup, has higher viscosity than relatively freely flowing water; you would notice more resistance when stirring syrup, for example, compared with water. Even air has viscosity, although under normal conditions it is about fifty times less viscous than water.

Viscosity arises because the movement of a molecule relative to its neighbouring molecules is resisted by the intermolecular forces between them. Viscosity has important consequences for flow in the vicinity of surfaces. At the very interface between a surface and a fluid, because of the attraction force between the molecules or atoms of the fluid and those of the surface, the relative speed between the surface and the fluid is zero. So, for a viscous fluid, the speed of the fluid relative to the surface must change from whatever its value is far away from the surface to zero, as seen in Figure 4.11. This layer where there is a strong velocity gradient is called the *boundary layer*.

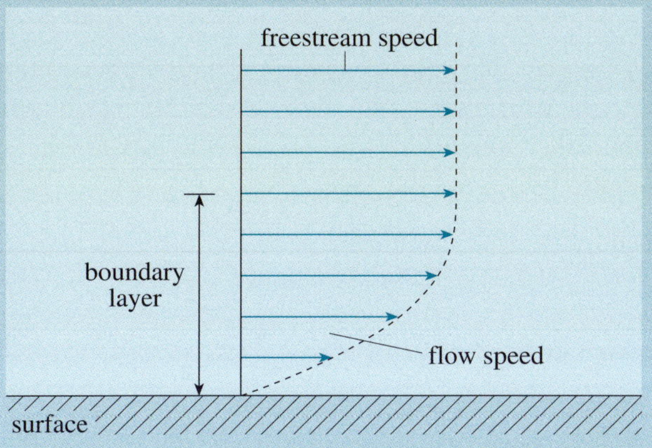

Figure 4.11 Boundary layer of a viscous fluid flowing over a solid surface

The thickness of this layer is directly proportional to the viscosity of the fluid. Although the layer is usually very thin as far as the aerodynamics of aeroplanes is concerned (for example, for the wings of a jumbo jet in cruising flight it is only a few centimetres thick), aerodynamic forces are greatly influenced by it. For instance, one component of drag is a skin-friction arising from viscosity: energy losses occur due to friction between adjacent air layers within the boundary layer (analogous to friction between solid surfaces).

A perfect or ideal fluid would be free from these complications, and it is defined as an imaginary fluid of constant density (incompressible) and zero viscosity (inviscid). 'Ideal fluid' must not be confused with 'ideal gas'. An 'ideal gas' is ideal in the sense that it conforms to certain gas laws and therefore it is compressible.

Returning to the streamline patterns in Figure 4.9, you can see that the streamlines above both the inclined flat plate and aerofoil are more closely spaced relative to the streamlines below them. That means that air flows faster where the pressure is supposed to be lower (at least that would be consistent with lift). The relation between flow velocity and pressure in ideal fluid flow was first formulated by Daniel Bernoulli in 1738. He based his analysis on the *conservation of energy* principle as applied to an incompressible fluid. That is, for an ideal fluid along a streamline the total energy is constant:

$$\text{kinetic energy} + \text{potential energy} = \text{constant}$$

You can anticipate that the kinetic energy will involve the mass of the fluid and the square of its flow speed. The potential energy needs careful thought. We need to include potential energy stored in the *pressure* of the fluid along with gravitational potential energy. The resulting equation gives one of the most important fundamental relations in fluid dynamics. In aerodynamics we are chiefly concerned with the interplay between airflow and air pressure around an aerofoil. The small changes in gravitational potential energy are unimportant, so the conservation of energy for a unit volume of airflow round a wing is:

$$\tfrac{1}{2}\rho v^2 + p = \text{constant}$$

where p is the pressure, v is the flow speed and ρ is the density of the air.

The units of p are $\text{N m}^{-2} = (\text{N m})\ \text{m}^{-3} = \text{J m}^{-3}$. $\tfrac{1}{2}mv^2$ is measured in J, so $\tfrac{1}{2}\rho v^2$ must be in J m^{-3}.

In fact, p is referred to as the *static pressure* and the quantity $\frac{1}{2}\rho v^2$ is referred to as the *dynamic pressure*, representing the kinetic energy of the flow. This is the simplest form of *Bernoulli's equation*, which states that in a streamline fluid flow, the sum of dynamic and static pressures remains the same; as one increases, the other must decrease. Hence, the greater the speed of the flow, the less the static pressure; and vice versa.

Air is not an ideal fluid, though. It is compressible at the cruising speeds of big airliners and it has appreciable viscosity. The actual airflow over an aerofoil is slightly different from what Bernoulli suggests, with a consequent reduction in lift and increase in drag. But it gives the basic idea.

3.4 So, how is lift generated?

Why, in the first place, does air flow faster over the top? Is it that 'because the pressure is lower, the air flows faster' or is it that 'because the air flows faster, the pressure is lower'? Bernoulli's principle does not answer this – it simply relates flow speed to pressure for an ideal fluid.

Some scientists have recently proposed an alternative explanation for lift generation. As in the *Newtonian lift* that I described earlier, their theory is also based on Newton's Third Law; the wings push the air down and, in return, the deflected air pushes the wings up (i.e. the lift is a reaction force). The credibility of this theory relies upon the explanation of how the wings divert sufficient amount of air downwards. This was the shortcoming of the Newtonian description of lift, where the air particles were assumed to deflect by bouncing off the bottom of the wing; the top surface of the wing was considered not to take any part in this process. But, if you study the streamline patterns in Figures 4.9(a) and (b), you will notice that the air flowing over the wing seems to be bending, following the contour of the surface. The reason for this behaviour of air is its viscosity, which gives it a kind of 'stickiness'. You may have noticed this effect at the kitchen sink; if you hold a spoon vertically under a stream of running water such that the water just touches the back of the spoon, the water tends to follow the spoon's surface rather than flowing straight down. This is known as the 'Coanda effect', after Romanian physicist Henri-Marie Coanda. It is claimed, therefore, that aeroplane wings act like giant 'spoons', diverting air downwards.

After a century of manned flight, scientists and engineers still debate how to explain the way wings produce lift. Fortunately, it is not necessary to understand every last detail of the physical principles of lift generation to quantify the forces and hence design an efficient wing.

3.5 Quantifying lift and drag

3.5.1 Lift

What factors are likely to affect the amount of lift experienced by a given aerofoil? Wind tunnels and intuition can help answer that question by looking at more specific issues.

Does the air speed matter? Yes: the faster the relative speed between the air and a wing, the greater the lift. Wind tunnel tests show that lift increases with the square of the air speed.

Does the density of the air influence lift? Yes: the denser the fluid, the greater the lift – after all, if the air were pumped out its density would approach zero and there would be nothing to promote lift.

Do the shape of the wing and its angle of attack matter? Yes: different shapes and different angles of attack, i.e. wing geometry, directly affect lift. That is one of the main things shown by experiments in wind tunnels.

We can distinguish between factors associated with the air and factors associated with the geometry of the aerofoil with the following empirical equation for the lift experienced by a wing based on these observations:

$$F_L = \left(\tfrac{1}{2}\rho v^2\right) \times \left(C_L \times A\right) \tag{4.3}$$

This is the 'lift equation' for the lift force F_L.

You can see the density and speed factors in the first group. The factor of ½ is included because it then looks like a term in Bernoulli's equation. In the second term, C_L, the *lift coefficient*, contains the geometric factors that describe the extent to which the lift force depends on the shape and setting of the wing. The wing area, effectively the area seen in plan view (from above), is A. Wind tunnel measurements are the simplest way to determine the lift coefficient, provided the wing has already been built – it is a property of the wing shape and the angle of attack (Figure 4.12).

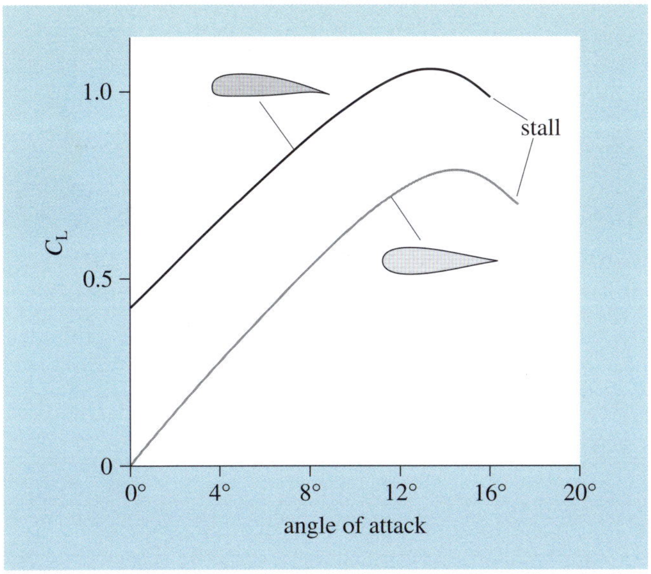

Figure 4.12 Lift coefficient (C_L) plotted against angle of attack for two different wing shapes

---

Exercise 4.4

For take-off and initial climb of large airliners the area of the wing is extended by sliding out extra plates from pockets along the trailing edge of the wing. This is aimed at getting more lift from the wing geometry by increasing the wing area (A in the lift equation). Take-off speed is then that speed for which the lift force just exceeds the aeroplane's weight.

Why then is the wing area similarly extended when large airliners come in to land?

Wind tunnels show that as the angle of attack increases, so does the lift coefficient, up to some critical point where suddenly the lift drops away to zero. This sudden loss of lift is called *stall*. It occurs when an aerofoil is no longer effective in 'spooning' the flow downwards. To recover lift, the angle of attack must be lessened.

3.5.2 Drag

Lift is just one component of the force experienced by a wing moving through the air. The other component is drag, which has exactly the same dependence on density, speed and area as the lift. A *drag coefficient* (C_D) expressses how the drag force scales with wing shape and the angle of attack (Figure 4.13). The drag equation looks very similar to that for lift:

$$F_D = \left(\tfrac{1}{2}\rho v^2\right) \times \left(C_D \times A\right) \tag{4.4}$$

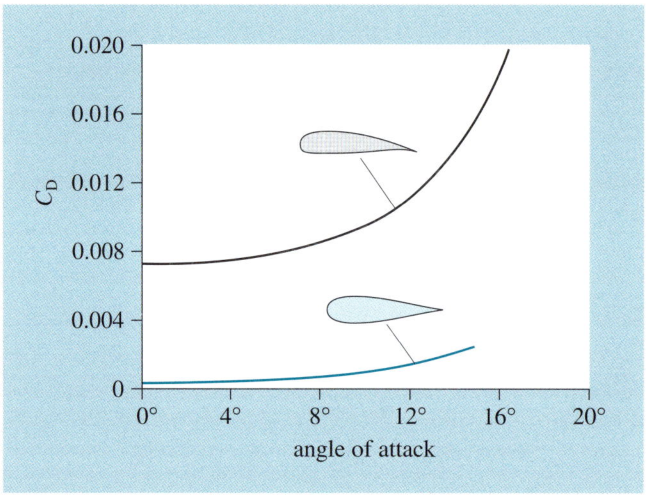

Figure 4.13 Drag coefficient (C_D) plotted against angle of attack for two different wing shapes

The chief challenge for the designers of wings is to find shapes that give higher lift and less drag. In other words, they seek to maximize the ratio of lift to drag forces F_L/F_D. This amounts to maximizing the ratio C_L/C_D and that is what wing design is all about.

We have already encountered viscous drag in Part 2 of this block, when we were working out the effect of air resistance on the performance of an accelerating car. Drag is more important for aeroplanes than for cars because the speeds are considerably higher (remember, drag forces scale with the square of the speed).

There are two contributions to drag forces. One is a skin friction arising from the shearing of flow in boundary layers. The other comes from the distribution of pressure around an object in a flowing fluid.

The skin friction and pressure drag on an aeroplane associated with generating lift is called *induced drag*. There is also additional skin friction and pressure drag caused by airflow over all the other parts of the aeroplane (fuselage, engines, etc.), and this too resists forward motion. This is called *parasitic drag*.

3.5.3 Factors affecting lift and drag

The amount of lift and drag generated by the wings is dependent on:

- the angle at which the wings meet the airflow or flight path;
- the shape of the wings, particularly in cross section – the aerofoil;
- the density (i.e. mass per unit volume) of the air;
- the speed of the airflow;
- the wing surface area.

3.6 Engine options

Thrust is the force produced in the direction of motion. It is required to overcome drag and thus to sustain the forward motion of the aeroplane, thereby generating lift. How do you produce this force up in the air? Well, we need turn to our maestro Newton again. His Third Law states that every action produces a reaction, equal in magnitude and opposite in direction. What we need is something to push against. There is plenty of air, and we know that fluids also obey Newton's law of mechanics (see **▼Air is a fluid▲** earlier in this part). Indeed, thrust is produced by the reaction to pushing air backwards by mechanical means through the engines of the aeroplane. The engine draws in a large mass of air at the front and thrusts it backwards.

There are two basic propulsion systems: propeller-produced and jet-produced. For a propeller-powered aeroplane, the air is accelerated upon passing through the rotating propeller. The blades act like rotating wings; the cross-section of each blade has an aerofoil shape. Thrust is obtained because the blade has a small inclination to its direction of motion, just like the angle of attack of an aerofoil. The details are rather complex but, put simply, the aerodynamics of the rotating blades effectively create a *lift* force in the direction of flight by gathering in and accelerating entrained air backwards.

Jet propulsion differs from propeller propulsion in the way it accelerates the air (or any gas). It usually involves expanding the gas through rapid heating (for example, by the combustion of suitable fuel) and ejecting the expanded gas from the engine at high speed. The chemical energy of the fuel is converted into potential (internal) energy in the form of an increase in gas pressure, which is converted into mechanical energy (work) upon its release from the engine as accelerated gas. Hero's aeoliphile (the engine shown in Figure 3.3 of Block 3 Part 3) is probably the first known example of jet reaction. The heat in a water chamber boils the water and the resulting water vapour (steam), which is at elevated pressure, escapes through the nozzles of the chamber. The jets of steam running through the nozzles, pointing in opposite directions, impart equal and opposite reactions to the nozzles, thus causing the chamber (the engine) to revolve.

The development of aeroplanes runs side by side with the advance in engine technology. Before the 1950s, the propellers were turned using a reciprocating or piston engine, which transmits the power to a shaft connected to the propeller through a piston and crank. During the Second World War, a new engine called a *gas turbine* engine (popularly known as the *jet engine*) was developed by British engineer Frank Whittle. Like the reciprocating engine, the gas turbine is also an internal combustion engine. Air is compressed in a chamber, mixed with a suitable fuel (commonly

Clearly, within jet engines, the idea of air being incompressible needs to be revised. In a jet engine, air is squeezed so hard it does get compressed.

kerosene), the mixture is ignited and the rapid expansion of the hot gas produces the power.

Unlike in the piston engine, the combustion in a gas turbine engine is continuous and the power is a result of forcing the expanding gas out of the rear of the engine. There are many variations of jet engines but they all work in a similar way. Figure 4.14 shows a type that is popular in civil aviation. At its heart are the three main parts of a gas turbine engine: compressor (sets of blades attached to rotating disks), combustion chamber and turbine (again, stages of rotating blades). Air is drawn in by the compressors and pressurized before arriving in the combustion chamber where it is mixed with vaporized fuel and the mixture is ignited. The expanding hot gases leave the chamber through the turbine section and out of the rear of the engine. The turbines are attached to the compressor disks through one or more concentric shafts. Part of the jet's energy is used to rotate the turbines, and then transmitted through the shafts to drive the compressors. The rest provides thrust from the expelled gases.

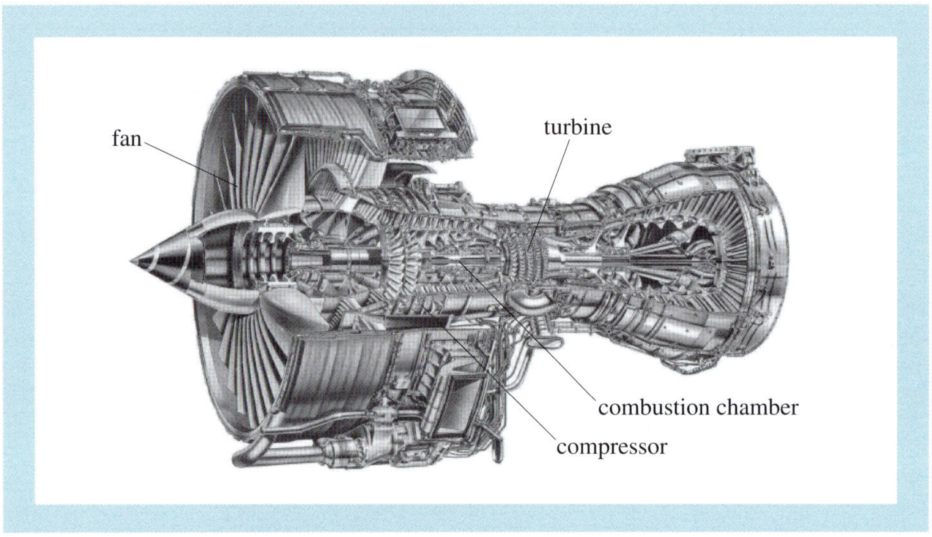

Figure 4.14 A turbofan engine

The gases pushed out of a jet engine and the air swept back by a propeller amount to a flow of mass. Whenever mass moves, there is momentum associated with the motion. **▼Keeping track of momentum▲** is the way to link the momentum flow to thrust.

The same thrust could be provided by a large mass-flow rate and small speed, or by a small mass-flow rate and large speed. In other words, pushing a large volume of air at a small speed is equivalent to pushing a relatively smaller volume of air at a higher speed. This is the main difference between propeller propulsion and jet propulsion. A propeller displaces a large volume of air at a moderate speed, whereas a jet engine forces out a smaller volume at a much higher speed. This difference has significant implications for the fuel efficiency of the engine.

▼Keeping track of momentum▲

From earlier studies of dynamics, you should know that one way to view forces on moving objects is as the rate of change of momentum. For propellers and jet engines, the forces they exert on a moving fluid can be traced back to changes in the rate of momentum. Notice the different order of the words – for moving solids it is convenient to look at the change of momentum with time, whereas with moving fluids it is simpler to look at how the rate of momentum flowing with the fluid changes from one place to another.

Figure 4.15 shows streamlines that trace fluid flow past a propeller. What does the squashing together of the streamlines suggest?

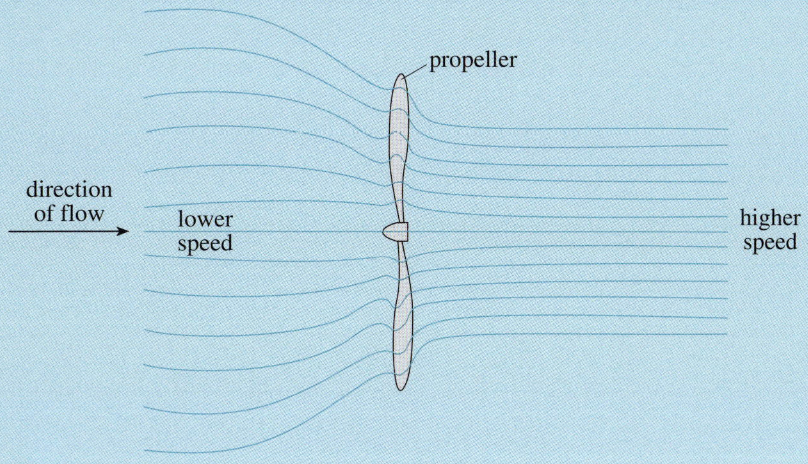

Figure 4.15 Streamlines past a propeller

The flow is faster where streamlines are more closely bunched. So the flow speed is increased as fluid passes the propeller. The flow on to the propeller of say 10 kg s^{-1} of air at a particular speed is squeezed down to a narrower, faster flow of 10 kg s^{-1} away from the propeller. Using the symbol $\dot{m}$ to denote the rate of mass flow, I can write the rate of flow of momentum on to the propeller at incoming speed v_i as:

incoming momentum flow rate $= \dot{m}_i v_i$

Similarly, downstream of the propeller:

outgoing momentum flow rate $= \dot{m}_o v_o$

Since the mass of air flowing in and the mass flowing out in each instant of time must be the same:

$$\dot{m}_i = \dot{m}_o = \dot{m}_{air}$$

So in passing the propeller, the change in the rate of flow of momentum, which I have identified as a force, is:

$$F = \dot{m}_{air}(v_o - v_i)$$

That is the force exerted by the propeller on the flow. Newton would only be happy if an equal and opposite force is simultaneously experienced by the propeller and engine. So there's the thrust. To get more thrust you need bigger propellers to gather in more air or faster propellers, both of which will increase the mass flow rate, or more twisted propellers to squash the air down to a narrower, faster exit stream.

The next exercise encourages you to work out how the model for thrust must be changed to cover gas turbine jet engines.

Exercise 4.5

In a gas turbine jet engine, thrust is obtained from the expulsion of exhaust from the combustion of fuel.

(a) What is the origin of the increased speed of the exhaust stream?

(b) In addition to the mass flow rate of intake air, what other mass flow needs to be taken into account?

(c) Write an equation for the thrust F_T produced by a gas turbine jet engine.

It turns out that propellers are capable of much greater efficiency than jets and you must be wondering why modern passenger aeroplanes do not use them any more. Well, the problem is that with increasing flight speed the speed of airflow at the tip of propeller blades exceeds the speed of sound (1 Mach, or 'Mach one'), which causes the efficiency of the propeller to drop drastically. Jet propulsion can work better at speeds above 0.5 Mach. Furthermore, jet propulsion enables flight at higher altitude.

The modern jet engine uses the best of both systems. A fan placed in front of the compressors brings some of the propeller efficiency to the gas turbine engine. The turbines drive the fan, as well as the compressors, and a large volume of air passes around the core section of the gas turbine engine (see Figure 4.14). Thus, a larger volume of air is displaced at a lower speed, in addition to the jet through the core section of the engine. The ratio of 'bypassing' air to the air going through the core section of the engine is about 8 for engines of large commercial aeroplanes. Although the limiting flight speed for this engine is lower than the turbojet, it is perfectly adequate for common cruising speeds around 0.8 Mach. Reduced noise level is a key advantage of *turbofans* over *turbojets*. Whether bypassed or not, the construction of jet engines is almost as specialized as that of the airframes. The engineering principles are fairly sound but finding **▼Materials for gas turbines▲** is a major challenge.

The main advantage of a gas turbine engine is its superior power-to-weight ratio over the piston engine. For this reason, some propeller-powered aeroplanes use gas turbine engines to drive the propellers. This propulsion system is called *turboprop*, which is a turbojet with an additional turbine stage to drive the propeller through a shaft and an adequate reduction gear. Almost all the thrust of the turboprop comes from the propeller.

▼Materials for gas turbines▲

There has been a considerable challenge to develop suitable materials and designs that can endure the adverse conditions of high temperatures, a corrosive environment and high stresses that occur deep inside a jet engine. However, the potential rewards through improved efficiency and performance are high.

The fuel efficiency and the performance (i.e. power-to-weight ratio) of gas turbine engines, like all internal combustion engines, are largely dependent on two factors: the compression ratio of the gas and the temperature of the hot gas as it leaves the combustor (i.e. the turbine inlet temperature). The larger the amount of compressed gas at higher temperature, the more gas will be expanded to a higher speed and hence the more thrust will be available.

In a typical turbofan engine used in large jets, the compressors can increase the pressure of the air by a factor of more than 30, and the temperature of the air can reach over 500 °C in the compression process. The compressed air is then mixed with fuel and burnt in the combustor, where the flame temperature can reach up to 2000 °C. The expanded hot gas then goes through the turbine section; the turbine inlet temperature can be as high as 1500 °C. The rotation of the turbine blades depends on the design of the engine (e.g. temperatures, blade sizes, etc.), but may be higher than 10 000 rpm (revolutions per minute).

Key factors for the efficiency of a gas turbine engine are the endurance limits of the materials used, particularly in the first turbine stage nearest to the combustor. Here a turbine blade, stressed by high-speed rotation, is required to retain its strength and dimensions at high temperatures of over halfway to the melting temperature of the most refractory metals. Under these conditions, materials may stretch and eventually break. The turbine material also has to have sufficient toughness (against possible impacts from birds and airborne debris) and it has to be resistant to fatigue, corrosion, vibration and wear. Although certain ceramics have high melting temperature, high strength and low density compared to most metals, so far their brittle nature has prevented them from being used in turbines.

'Creep' of materials will be discussed in Block 5.

The current choice is nickel-based alloys which retain their strength at temperatures just above 1000 °C, but the aim is to operate them at above 1400 °C. The blades made of these alloys can only withstand the extreme conditions by protecting them with complex coatings and extensively cooling them by passing air through internal passages. Two types of coating are used; first a corrosion-resistant coating (based on nickel, chromium, aluminium, etc.) directly on the bare alloy, and onto this a ceramic coating (e.g. zirconia) to provide an insulating barrier to extreme heat (thermal barrier coating). The total coating thickness is about half a millimetre. A similar alloy is also used in later stages of the turbine, but as the temperature is not as high, internal cooling is not required.

The material choice for compressor blades is usually a titanium alloy, which has a temperature capability of around 600 °C. Why titanium?

Titanium has a higher strength per unit mass (σ_y/ρ) than nickel – every kilogram saved in engine mass without loss of performance could be part of a fare-paying passenger. You can be sure that alternatives are being actively sought that perform better and cost less.

Titanium alloy is also the preferred choice for the fan blades of a turbofan engine. The fan (visible from the front of the engine) in a large engine is up to 3 metres in diameter and operates at ambient temperatures (i.e. up to 50 °C at ground level and about −50 °C at an altitude of 12 000 m). It can pump more than half a tonne of air per second. Stiff, lightweight blades are manufactured by bonding two skins around a honeycomb structure. The resulting blade is less prone to vibrations due to uneven airflow, and is lighter than a solid blade.

Example

Let us calculate the stress, due to rotational force, at the root of a fan blade made of titanium (density 7200 kg m^{-3}) rotating at 3000 rpm (Figure 4.16). Assume that the cross-sectional area of the blade is constant throughout its length. The distances from the root and the tip of the blade to the centre of rotation (i.e. centre of the disc to which the blades are attached) are 200 mm (r_i) and 1500 mm (r_o), respectively (giving the blade length as 1300 mm). The tensile strength of titanium is 1000 MN m^{-2}.

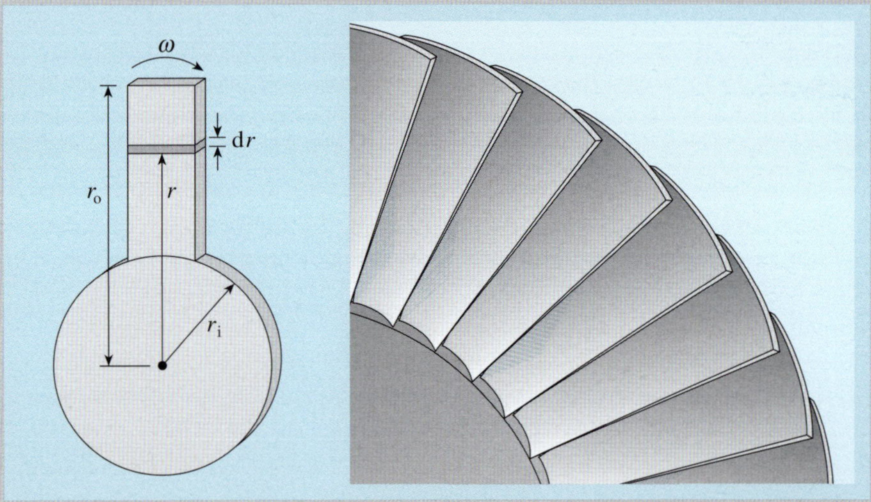

Figure 4.16 A fan blade – rotational stress calculation

Solution

For a mass m to follow a circular path of a radius of r with an angular speed ω (rad/s) there must be a centripetal force:

$$F = m\omega^2 r$$

For a blade rotating as shown in Figure 4.16, between the root and the tip of the blade the required force increases with the distance from the centre of rotation. The maximum force is at the root of the blade, since the root holds on to the rest of the blade.

We can calculate this force by imagining the blade is made up of small elements with a thickness of dr at a distance r from the centre of rotation. If we add up the forces caused by every element at a distance r varying from the root to the tip, we can determine the force at the root. The accuracy of the process is dependent on the choice of element thickness, dr; the smaller it is, the more accurate the result gets. Mathematically, it is possible to get an exact solution using integration. This is what I am going to use. First, let us calculate the force required to keep the small element whose mass is dm moving in a circular path:

$$\mathrm{d}F = \mathrm{d}m\ \omega^2 r$$

Note that the rotational speed ω is constant through the length of the blade. The mass of the element is:

$$\mathrm{d}m = \rho A\,\mathrm{d}r$$

where ρ is the density of the material and A is the cross-sectional area. Substituting gives:

$$\mathrm{d}F = \rho A\,\mathrm{d}r\ \omega^2 r$$

Integrating both sides:

$$\int_0^{F_{\text{total}}} \mathrm{d}F = \int_{r_\text{i}}^{r_\text{o}} \rho A\omega^2 r\ \mathrm{d}r$$

Since ρ, A and ω don't change with radius they can be taken outside the integration:

$$\int_0^{F_{\text{total}}} \mathrm{d}F = \rho A\omega^2 \int_{r_\text{i}}^{r_\text{o}} r\ \mathrm{d}r$$

which gives:

$$F_{\text{total}} = \rho A\omega^2\left[\frac{r^2}{2}\right]_{r_\text{i}}^{r_\text{o}} = \tfrac{1}{2}\rho A\omega^2\left(r_\text{o}^2 - r_\text{i}^2\right)$$

The stress at the root is:

$$\sigma = \frac{F_{\text{total}}}{A}$$
$$= \tfrac{1}{2}\rho\omega^2\left(r_\text{o}^2 - r_\text{i}^2\right)$$

Note that the stress due to rotation is a function of the density of the blade material, the rotational speed and the dimensions of the blade and the disc. Putting in the values gives:

$$\sigma = 7200\ \text{kg m}^{-3} \times (3000\ \text{rpm} \times 2\pi/60\ \text{rad s}^{-1})^2\ (1.5^2 - 0.2^2)\ \text{m}^2/2$$

$$\sigma = 785 \times 10^6\ \text{kg m}^{-1}\ \text{s}^{-2} = 785 \times 10^6\ \text{N m}^{-2}$$

Comment: The stress at the root is almost 80% of the tensile strength. I hope no one will risk a 10% increase in rotational speed without a major redesign.

3.7 Stability and control

Can we fly our aeroplane now, since we have sorted out the lift required to overcome the weight of the aeroplane, and the thrust provided by the engines to defeat the drag forces? Not just yet! We have to give the pilot the necessary means to control the aeroplane and the ability to manoeuvre it in three-dimensional space.

Exercise 4.6

Describe the degrees of freedom of movement of the following transport vehicles: a train, a car, a bicycle and an aeroplane.

The movement or manoeuvring of an aeroplane in any way about three axes is achieved by manipulating the moments of aerodynamic forces about the centre of mass of the aeroplane – see **▼Centre of pressure▲**. If the total moment about the centre of mass of the aeroplane is not zero, the aeroplane will rotate in accordance with the resultant moment. That is, the manoeuvring of an aeroplane relies on imbalanced moments.

▼Centre of pressure ▲

The aerodynamic force generated by a moving object is a result of the pressure difference around the object. Figure 4.17 shows how the pressure varies around a section of the wing (aerofoil).

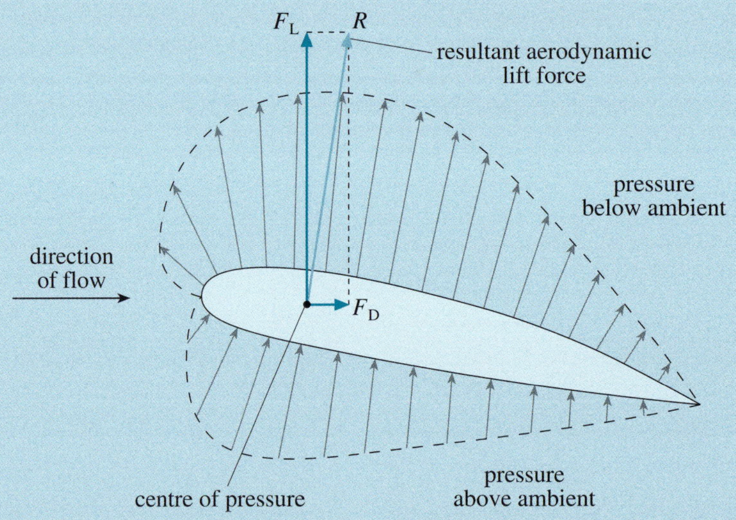

Figure 4.17 Pressure variation over an aerofoil. The centre of pressure is the average location of the total pressure acting on the wing

Since the pressure distribution is not uniform, the magnitude and direction of the 'net' or resultant aerodynamic force are found by summing the forces generated at every point on the surface of the aerofoil (that is, mathematically speaking, by *integration*). In effect, this is done by dividing the surface area into sufficiently small units such that the pressure in that unit may be taken to be constant. The force in each unit is then calculated by multiplying the pressure on the unit by the area of the unit (force = pressure × area). The summation of all the force vectors will give the magnitude, direction and location of the resultant aerodynamic force. This single force can be considered to act through the average location of the pressure on the surface of the aerofoil. This average location is called the *centre of pressure* (or centre of lift). Resolving the aerodynamic force into components in a direction perpendicular and parallel to the airflow gives the lift and drag forces, respectively.

The idea of centre of pressure is similar to *centre of mass*, which can be visualized as the point at which the whole mass of a body acts.

The relative position of the centre of pressure and centre of mass of the whole aeroplane is therefore crucial for longitudinal stability of the aeroplane; if they don't match up, the aeroplane will rotate.

The centre of pressure of an aerofoil is not fixed. Since the pressure distribution varies with the angle of attack, the position of the centre of pressure also varies (Figure 4.18). For conventional cambered aerofoils, it moves forwards with increasing angle of attack. Note that the magnitude and direction of the resultant force (and hence the amount of lift and the drag forces) also change.

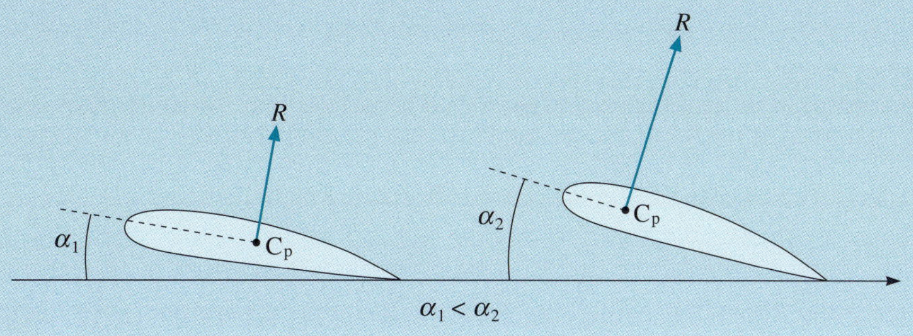

Figure 4.18 Centre of pressure and angle of attack

Before learning how to manoeuvre the aeroplane, we should find out if we could manage a level flight by balancing all the moments about the centre of mass of the aeroplane so the total moment becomes zero. When there is no rotation about the centre of mass, the aeroplane is said to be *trimmed*. Figure 4.19 shows the case for the moments created about the pitch axis throughout the centre of mass (turning the nose up or down) by the line of action of the four forces on an aeroplane. For clarity, I have drawn the pitching moment created by vertical forces (lift and weight) and horizontal forces (thrust and drag) separately; obviously, the total pitching moment is the combination of the two.

Taking moments about the centre of mass of the aeroplane gives the pitching moment. The moment about the centre of mass is the force times the perpendicular distance between the point of application of the force and the centre of mass (moment arm). Clearly, the weight of the aeroplane acts through the centre of mass, thus not contributing to the moment, so the pitching moment caused by the vertical forces is:

$$M_{\mathrm{Pv}} = F_{\mathrm{L}} d_{\mathrm{L}}$$

where d_{L} is the distance between the lift centre of pressure and the centre of mass.

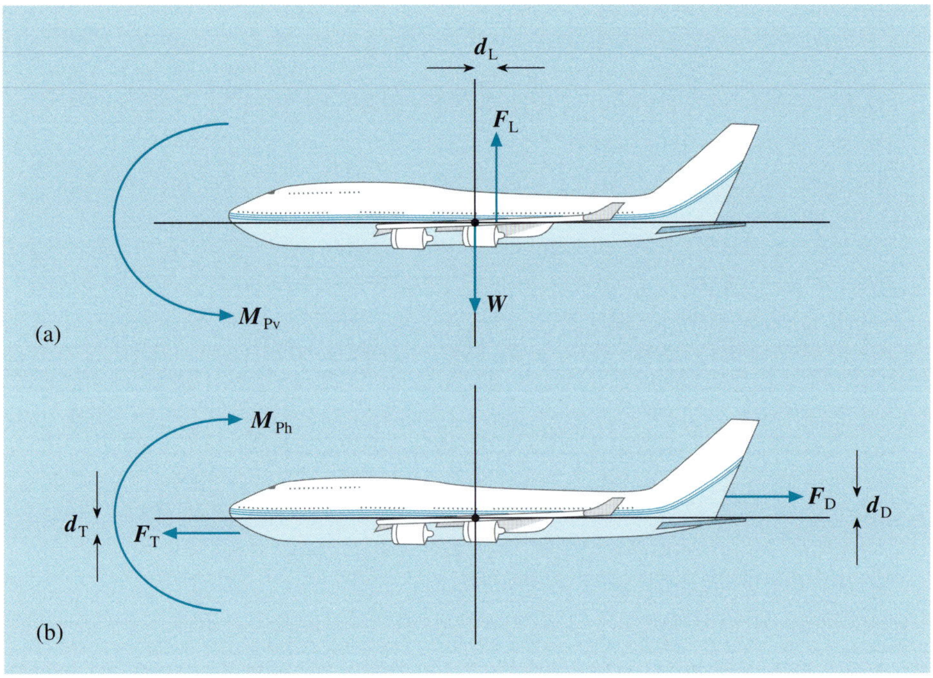

Figure 4.19 Pitching moment due to (a) centre of pressure of lift force behind the centre of mass, and (b) lines of thrust and drag forces not collinear. The total pitching moment is created by the combination of the two cases

Similarly, the pitching moment caused by the horizontal forces is:

$$M_{Ph} = F_T d_T + F_D d_D$$

where F_T is thrust, F_D is the drag force, and d_T and d_D are the perpendicular distances between the lines of action of the two forces and the centre of mass of the aeroplane. Note that, for the case shown in Figure 4.19, the lift causes a nose-down moment (a negative moment by sign convention) and both horizontal forces cause nose-up moments (positive moments). The total pitching moment is then:

$$M_P = F_T d_T + F_D d_D - F_L d_L$$

How can we make the pitching moment zero ($M_P = 0$)? We cannot alter the magnitudes of the forces since we have to maintain the force balance for given flight conditions. Varying the line of action of the forces (i.e. varying the moment arms d_L, d_T and d_D) is not possible either if wings and engines are fixed (though d_L does depend on where the angle of attack places the centre of pressure).

The conventional method to balance the pitching moment is to introduce an additional aerodynamic force generated by a secondary surface; the most common one is the *tailplane* (Figure 4.20). The main duty of the tailplane is to counterbalance the pitching moment created by the four basic forces. As it is located far from the centre of mass of the aeroplane it has a high leverage, so that it does not need to generate a large force to trim the turning moment to zero. Because the trimming of moment is done with a small force the total lift of the aeroplane is not significantly disturbed.

SAQ 4.3 (Outcome 4.5)

Why is it beneficial to have the tailplane positioned as far as possible from the wing in the design of an aeroplane?

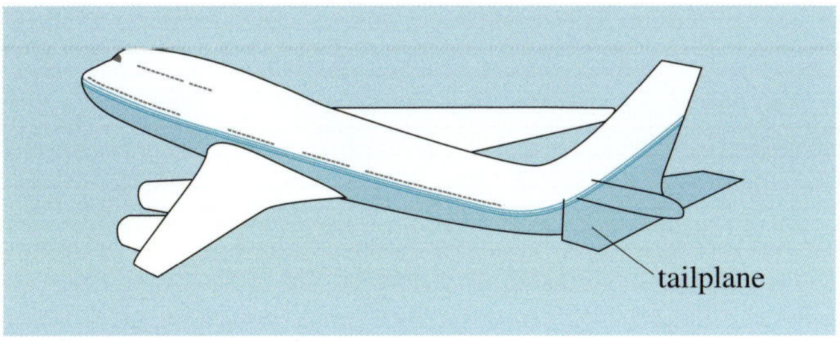

Figure 4.20 The tailplane

Exercise 4.7

The engines on large jet airliners are usually located under the wings. This gives the line of action of the thrust located below the centre of mass of the aeroplane. Comment on the benefit of this configuration in case of engine failure.

3.7.1 Stability

Why not, you may ask, design the aeroplane such that the force required from the tailplane is always upwards? Then the tailplane works with the wings by providing part of the lift required, as well as doing its main duty of balancing the pitching moment, as shown in Figure 4.21. I assume here that the lines of action of thrust and drag go through the centre of mass and therefore do not contribute to the pitching moment.

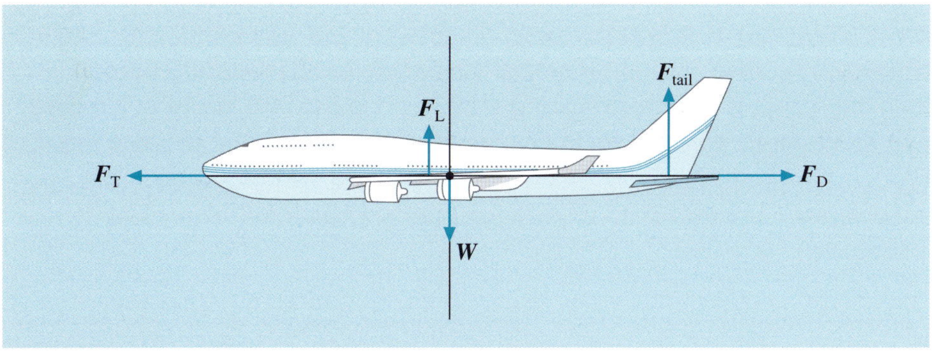

Figure 4.21 Balancing the pitching moments

It turns out that this configuration is not very stable. Planes frequently fly through disturbances in air – sudden gusts and turbulence – that may upset the equilibrium position. It would be very tiring for the pilot to continuously correct the balance of the aeroplane by, for example, adjusting the angle of attack of the tailplane. Imagine, for the configuration in Figure 4.22, where the centre of pressure is ahead of the centre of mass, that a slight disturbance in air flow causes the effective angle of attack of the wings to increase by a few degrees. Since the lift is roughly proportional to the angle of attack, wing lift will increase. Furthermore, the centre of pressure of the wing will move forwards, making the moment arm between centre of pressure and the centre of mass of the aeroplane larger. As a result, the nose-up pitching moment due to the wing lift increases, requiring even more correcting moment from the tailplane. Although the angle of attack of the tailplane increases as well, the increase in tail lift cannot cope with the requirement to keep the pitching balance. Therefore, unless the pilot corrects the pitching balance by increasing the tailplane angle of attack even further, there will be an ever-increasing nose-up pitching moment.

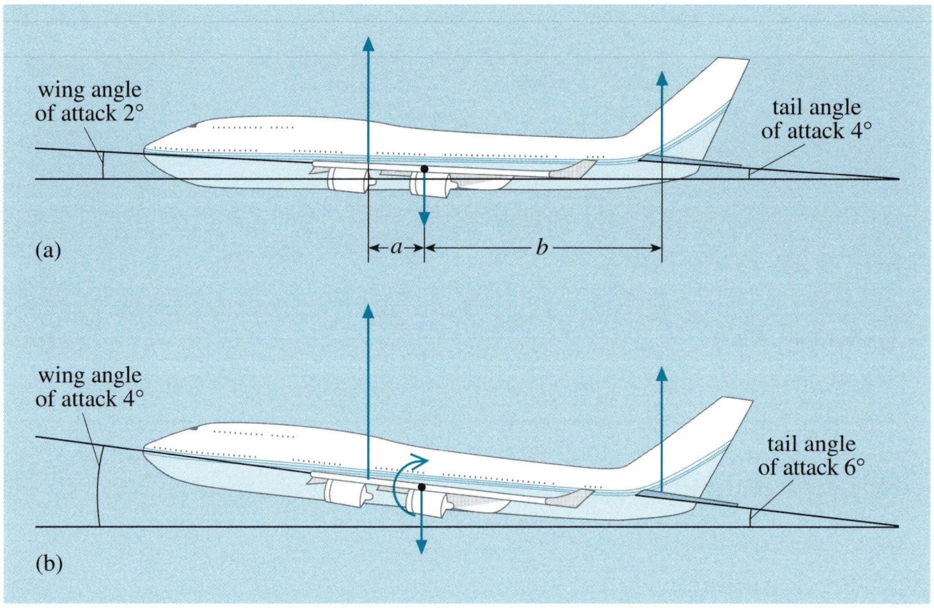

Figure 4.22 Pitching moments: centre of pressure ahead of centre of mass (angles of attack are exaggerated)

Now consider the configuration given in Figure 4.23, where the centre of mass is ahead of the centre of pressure. This configuration requires a negative (downwards) 'lift' from the tailplane for trimmed flight, giving a nose-up pitching moment. Again, suppose the angle of attack of the wings increases by a few degrees due to a slight upset in the air. Now, the angle of attack of the tailplane will decrease, resulting in less downward force and hence less nose-up pitching moment. The angle of attack of the wings, on the other hand, will increase giving a higher lift. But, the centre of pressure will move forwards, towards the centre of mass, making the moment arm shorter. As a result, the total pitching moment will be nose-down, causing the aeroplane to go back to its original trimmed condition before the disturbance.

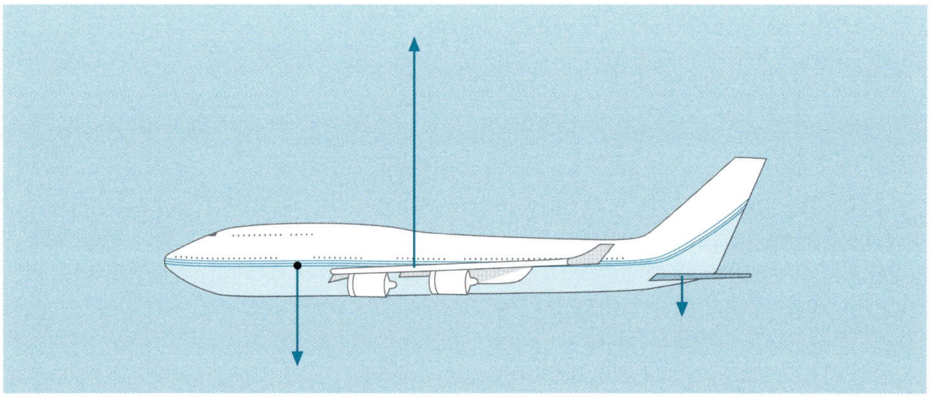

Figure 4.23 Pitching moments: centre of pressure behind centre of mass

This aeroplane is said to have longitudinal static stability, as it tends to pitch itself back into the equilibrium without the intervention of the pilot.

The directional or yawing stability (see Figure 4.33 in the answers to the exercises) is provided mainly by the fin (vertical stabilizer). Its working principle is similar to that of a weathercock. If the flight direction of the aeroplane is somehow upset, the airflow around the fin will produce a sideways force, and hence a yawing moment about the centre of mass of the aeroplane (Figure 4.24). This will tend to restore the flight direction automatically, without the pilot's intervention.

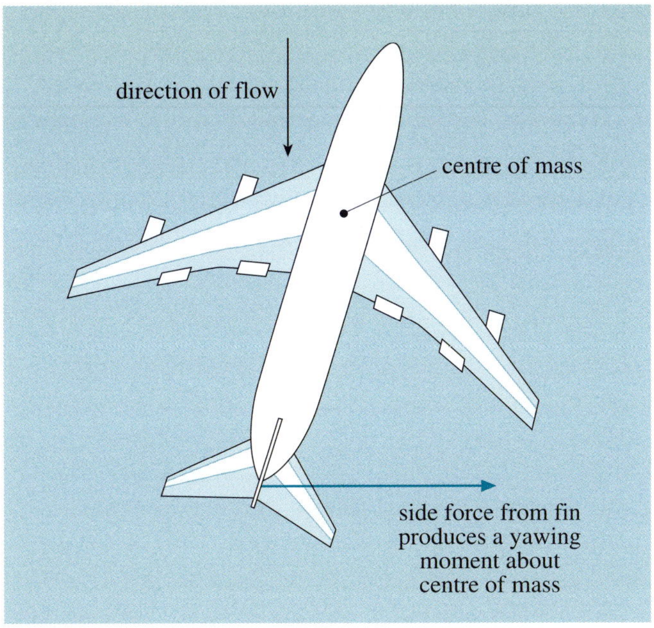

Figure 4.24 Directional or yawing stability

The most common method of achieving lateral or rolling stability is by the use of dihedral angle on the wing, as shown in Figure 4.25.

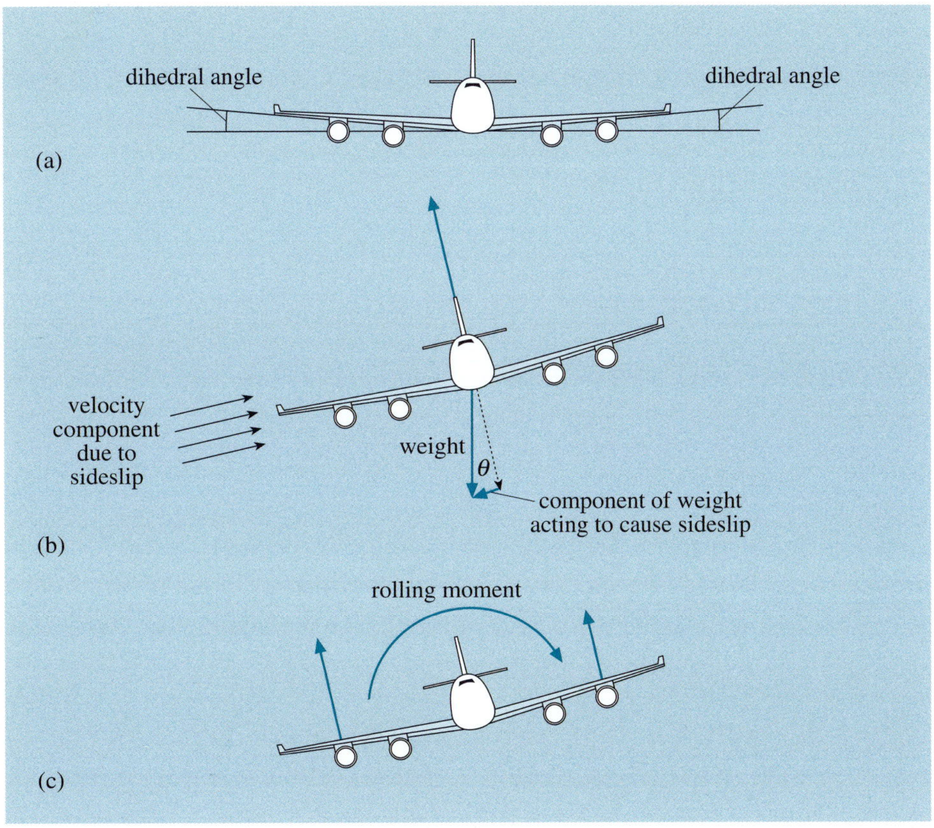

Figure 4.25 Lateral or rolling stability

If a disturbance causes one side to drop relative to the other, the resultant lift force will be inclined to the vertical, while the weight will obviously remain vertical. The two forces will not balance each other, and there will be a small resultant force acting sideways and downwards, which will cause the aeroplane to sideslip in this direction. This will result in a flow of air in the direction opposite to the slip. This has the effect of increasing the angle of attack of the lower side, and decreasing that of the higher wing. The lower side will therefore produce more lift, which will generate a rolling moment to restore the equilibrium position.

Although I presented them to you separately, yawing and rolling movements are interrelated; a roll will also result in yawing motion, and a yaw will cause rolling motion.

Stability of an aeroplane during flight is a very broad subject. The cases I have just introduced are simple ones that concern only the static stability, which alone is not sufficient to make an aeroplane stable. Figure 4.26 shows what forms of motion a statically stable plane may undergo during the restoration of pitching equilibrium. It may just restore its original position straight away, or more likely, it may overshoot a little, one way then the other, before eventually settling on the equilibrium position. This is a typical behaviour of a dynamically stable aeroplane. Figure 4.27 shows a dynamically unstable case, where the amplitude of the oscillations about the equilibrium position grows with time.

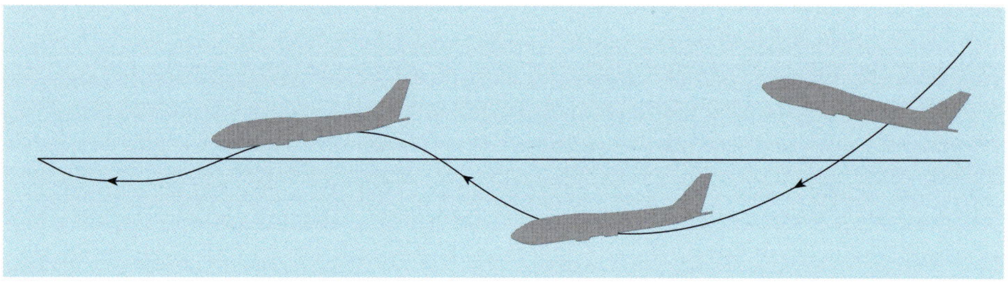

Figure 4.26 Stable pitching

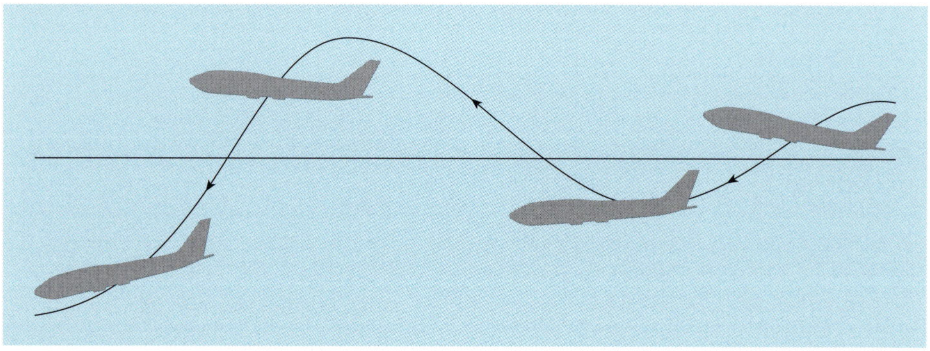

Figure 4.27 Unstable pitching

3.7.2 Control

Whether an aeroplane is stable or unstable, it is necessary for the pilot to have the means to change the aeroplane's flight conditions and manoeuvre it into any desired position. This is achieved by movable control surfaces, shown in Figure 4.28. The working principle is the same in each case: alter the aerodynamic force by moving the relevant surface, which will change the moment equilibrium and rotate the aeroplane about the required axis.

Elevators, rudder and ailerons are the principal movable surfaces providing the pitch, yaw and roll controls, respectively.

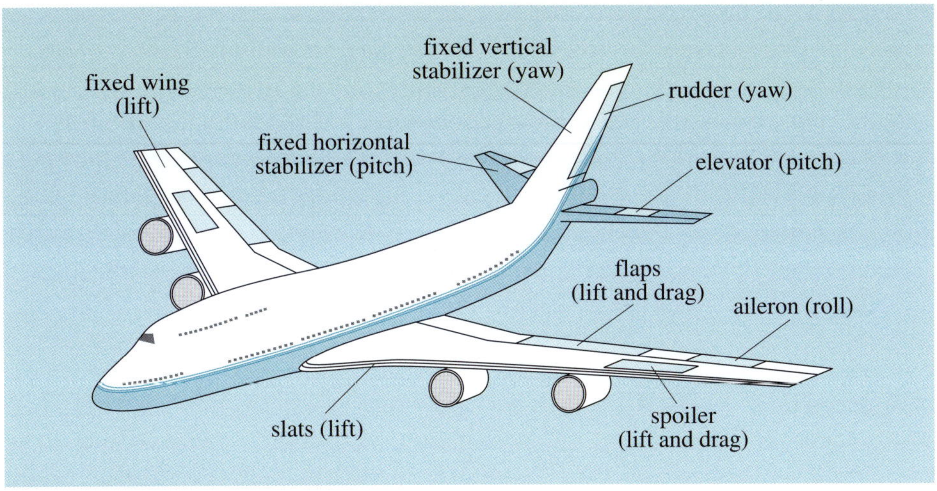

Figure 4.28 Control surfaces

In a conventional aeroplane, the rear portions of the tailplane are hinged to form the elevators. The tailplane itself is fixed into position and the movable elevators provide the means of momentarily adjusting the balance or manoeuvring the aeroplane. The shape of the tailplane is optimized to provide the balance of the aeroplane at cruising speed with the elevators in neutral position.

For example, to make the aeroplane roll, the ailerons in each wing are moved in opposite directions. Lowering the aileron on a wing increases the camber of the wing and hence increases the lift force, whereas raising an aileron does the opposite, i.e. reduces the lift. The resulting imbalance gives the aeroplane a rolling moment and the aeroplane rolls in the direction of the wing with raised aileron.

It seems easy to control the aeroplane! However, do not rush to get your pilot's licence yet. The interaction or cross-coupling between movements make the control complicated. The pilot will use a combination of controls to manoeuvre the aeroplane. For example, the ailerons and the rudder are used together to make a smooth banking turn. Furthermore, the design of control systems in large aeroplanes is much more complicated and, not surprisingly, **▼Computers take control▲**.

In addition to the primary control surfaces, that is ailerons, elevators and rudder, there are spoilers (small surfaces on the wings) which when activated spoil the flow over the wing and thus increase the drag and reduce the lift.

Stability and manoeuvrability are opposing characteristics. Highly stable aeroplanes can be very difficult to manoeuvre since the forces generated by the control surfaces have to overcome the stabilizing forces. On the other hand, too little stability means the pilot has to make frequent corrections to keep the aeroplane at the desired flight condition. The exact amount of stability built into an aeroplane depends on its role. Large transport planes are normally designed to have high stability, whereas planes that demand high agility, such as aerobatic planes and fighter jets, are inherently less stable.

▼Computers take control▲

In simple aeroplanes, the control surfaces are moved by the pilot using cockpit flight controls (such as stick, yoke and pedals) connected to the control surfaces with steel cables or push–pull rods. For light aeroplanes, the pilot's muscle power coupled with a leverage system is usually sufficient. However, in large or fast aeroplanes the aerodynamic forces are too great for the pilot alone to move the control surfaces and power assistance is provided in the form of hydraulic or electrical actuators. Powered by engine-driven pumps and a network of pipes, hydraulic actuators are a convenient way of providing the force required in operating many movable parts of the aeroplane, including control surfaces, retracting and extending undercarriage and wheel brakes, etc.

The flight control system where there is no mechanical connection between the flight controls and the control surfaces is known as 'fly-by-wire'. The movements of the flight controls (stick and pedals) are converted into electrical signals which are sent to the relevant actuators. The actuators then provide the necessary motion of the control surfaces.

The term fly-by-wire has been retained to describe a more sophisticated system that incorporates electronic circuitry and computers. In this system, sometimes referred to as digital fly-by-wire, the pilot's control inputs are fed into a computer (in practice, two or more identical computers) and are analysed together with information on current flight conditions supplied by many sensors around the aeroplane. The computer then checks the rules with which it has been programmed and decides which control surfaces to move and by how much. If the system considers that the pilot's command may be unsafe, it can dismiss it. Modern passenger jets use this system, which drastically improves the performance, efficiency and safety of the aeroplane. (The manufacturers claim that the computers can now fly the aeroplane with no need for a pilot, including take-off and landing.)

What I have described so far relates to the stability and control of a 'conventional' aeroplane; a cylindrical-tube-like fuselage to which is attached a pair of wings around the middle and a tail unit at the back, such as all commercial Boeing and Airbus transport jets. Aeroplane designers, however, are endlessly studying alternative designs for improving efficiency and/or performance. For example, some designs put the horizontal stabilizers towards the front of the fuselage, before the wings (tail-first or canard design), or designs with no tail at all. Some designs shrink the fuselage to reduce parasitic drag by making the wings wider and deeper (blended-wing design) or get rid of the fuselage altogether, making the aeroplane a flying wing. Nevertheless, as far as the stability and control are concerned, the basic principles remain the same: whatever your design is, make sure that the aerodynamic forces and the moments balance. For instance, for Concorde, the supersonic passenger jet, the shape of its wings was designed to provide the required lift at both subsonic and supersonic speeds. However, to reduce the drag at high speeds it did not have a tailplane. The stability of the aeroplane without the stabilizing effect of a tailplane is provided in an ingenious way (see **▼Concorde's tail▲**).

▼Concorde's tail ▲

Concorde's centre of pressure shifts about 2 m rearward at Mach 2.2, which gives it a nose-down pitching moment. It does not have a conventional tailplane to counterbalance this. The shape of Concorde is optimized to have a high lift-to-drag ratio at supersonic speeds; therefore, it was designed to exclude the conventional tail unit which would be very inefficient at supersonic speeds. Instead, Concorde has unique control surfaces called elevons at the trailing edges of its wing (Figure 4.29). Elevons have dual functionality: moving elevons on both sides in the same direction, up or down, controls the pitching moment (that is the function of elevators in a conventional aeroplane), and when they move in the opposite direction the aeroplane rolls about its longitudinal axis (i.e. they act like ailerons) – hence the name elevon (elevator plus aileron).

Figure 4.29 Concorde

However, balancing the pitching moment using elevons brings a hefty drag penalty, which is very undesirable for a supersonic flight. Concorde designers had a clever idea – why not change the centre of mass of the aeroplane in accordance with the shift in centre of pressure? How can you achieve this – you can't ask the passengers to move backwards and forwards all the time. An easier alternative is to shift the fuel forwards or rearwards. Any aeroplane carries a little more fuel than required to reach its destination. In Concorde, about 95 tonnes of fuel are carried in two groups of fuel tanks: the main group in the wings and fuselage, and the trim tanks at the front and rear of the fuselage. The aptly called trim tanks held about a third of the total fuel with two forward tanks and one rear tank located in the fuselage beneath the vertical tail fin. The transfer of fuel rearwards during aircraft acceleration to supersonic flight caused the centre of mass to move rearwards to match up with the movement of the centre of pressure. To achieve a 2 m shift in the centre of mass of the aeroplane, about 20 tons of fuel are displaced.

This idea also works for setting the aeroplane roll trim when, for example, flying through a side wind. The ability to move fuel between the main tanks on the wings allows the roll trim to be set without having slightly different deflection on elevons, and thus avoiding increased drag.

The idea of moving the centre of mass of an aeroplane in flight is not restricted to supersonic aircraft. It offers aircraft designers new options for controlling flight in conventional aeroplanes.

3.7.3 Turning

Finally, let's look into one of the basic manoeuvres in more detail: turning to change the heading of an aeroplane. Figure 4.30 shows the path that a turning aeroplane takes.

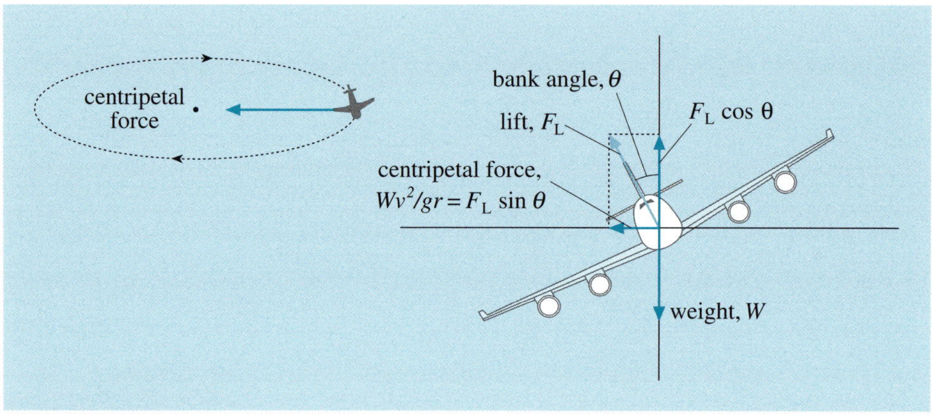

Figure 4.30 Forces in a banked turn

Exercise 4.8

Recalling the curvilinear motion that you studied in Part 1 of this Block, describe how an aeroplane can follow a circular path when turning.

Unlike ground vehicles which get their traction forces through the contact of their wheels with the road, aeroplanes must rely on aerodynamic forces for producing the centripetal force when following circular paths. As shown in Figure 4.30, the centripetal force is provided by the aeroplane banking, so that the total lift on the wings, in addition to providing a vertical component to balance the weight, can supply a horizontal component towards the centre of the turn.

If the aeroplane is banked at an angle θ to the horizontal and if the aeroplane is executing a proper level turn (i.e. it is not sideslipping and it is keeping its height constant) then the lift, F_L, will act normal to its wings. As you can see in Figure 4.30, the total lift required for level flight in a banking turn should be bigger than that in straight flight, since the vertical component of the lift must balance the weight of the aeroplane. Using the trigonometric relations that you ought be accustomed to by now, we can write an equation relating the bank angle, θ, to the required lift as:

$$mg = F_L \cos \theta$$

Since the cosine of an angle gets smaller with increasing angle, this equation tells us that with increasing banking angle the wings must provide more lift to keep the vertical balance. We know how to produce more lift; recall the lift equation (4.3):

$$F_L = C_L \times \left(\tfrac{1}{2}\rho v^2\right) \times A$$

The pilot has two choices: either increase the lift coefficient, C_L, by increasing the angle of attack or partially employing the high lift devices (flaps) or increase the speed, v.

The horizontal component of the aerodynamic lift of a banking aeroplane provides the centripetal force, mv^2/r. By using trigonometry, we can calculate the correct banking angle. Referring to Figure 4.30:

$$\tan\theta = \frac{v^2}{rg}$$

This equation tells us that the correct banking angle depends on the desired radius of turn and the speed of the aeroplane. It is interesting to note that this angle is independent of the aeroplane's weight: any aeroplane maintaining a speed v in level flight but banked at an angle θ will execute a path along a horizontal circle of radius:

$$r = \frac{v^2}{g\tan\theta} \qquad (4.5)$$

SAQ 4.4 (Outcome 4.5)

An aeroplane with a mass of 200 000 kg travelling at a speed of 400 km h^{-1} executes a gentle banking turn with a radius of 1000 m. If it maintains its speed during the manoeuvre, calculate the correct banking angle and the extra lift needed to produce a level turn.

4 ENGINEERING THE FUTURE OF CIVIL AIR TRANSPORT

Now that we know a little bit more about flight mechanics, can we, as engineers, help the manufacturers develop an aeroplane that they can sell in large numbers to the airlines?

As in any product, the economics is dominated by market forces: is there a demand for your product? Do we need a new aeroplane that looks like the others? The development of a large airliner requires vast capital investment, and the break-even period may be more than 10 to 15 years, depending on how many aeroplanes you can sell. The life of the design can be more than thirty years (the Boeing 747 has been around since the late 1960s, albeit with many design improvements and modifications). A new aeroplane is usually developed to utilize a new technology which brings certain competitive advantages to the airlines, such as more economy or speed.

So the question is, what kind of aeroplane will the airlines require in the future? No one knows the answer for sure. However, what we can do is to examine the important factors, such as capacity and speed, and try to come up with feasible options with the help of what we have learned so far about the aerodynamics of flight.

4.1 The business challenges

Listen to the audio track *Global Business – Up in the Air* (T207 CD-ROM) again, if necessary, and try to answer these questions which concentrate on the parts of the programme relevant to the future competition between Airbus and Boeing. It is important that you appreciate why design engineers in the two companies are being asked to solve different problems.

SAQ 4.5 (Outcomes 4.1, 4.2 and 4.6)

Question 1 – The investment

(a) What order of magnitude of investment do the two companies expect to make in the development of their new aircraft?

(b) How is this expenditure justified?

(c) What timescale is suggested as the period before either company will know whether its investment has succeeded?

Question 2 – The strategies

In no more than 100 words, outline the contrast between Boeing's and Airbus's strategies for their next-generation airliners. You should include comments on:

- anticipations of passenger expectations;
- differences in intended routes – with reasons;
- arguments for economical operation (what the airlines require).

Question 3 – Production systems

What is the major production-system difference between the two companies?

Question 4 – About the A380

(a) Was the design of A380 completed when the programme was made (2002)? When does it come into service?

(b) The A380 is described as 'huge'. How much more floor area per passenger (averaged across classes) does it offer than the existing Boeing 747-400? Do you think this justifies the hype proclaimed by John Lee?

(c) How does Airbus answer the Boeing view that 'point to point' flights are what people want?

Question 5 – About the Sonic Cruiser

(a) Is the Sonic Cruiser seen as a single aircraft type by Boeing? Has Boeing decided exactly what to build? When should Sonic Cruisers come into service?

In 2003 work on the Sonic Cruiser was suspended.

(b) Why is 'faster' better for the airlines as well as for the passengers?

(c) What has 'deregulation' led to in the US air transport market?

(d) Is there a logical limit to Boeing's concept of letting people fly the routes they want? (How much traffic would there be on the Timbuktu to Milton Keynes sector?)

(e) Sonic Cruisers will need to fly higher than present-day airliners. Explain why.

Question 6 – Forecast the outcome

The programme has tried to put the companies' strategies forward as a battle – 'bigger versus faster; who will win?'.

Summarize Daniel Solon's argument that both could be profitable.

4.2 The engineering challenges

The shape of the modern aeroplane has not changed for fifty years – a long tube with wings attached. However, advances in many fields have provided vast improvements in performance, of which the most important ones are: propulsion technology, aerodynamics, control systems (e.g. fly-by-wire) and structures and materials.

Among the engineering challenges for civil aviation are speed, capacity and operational costs.

4.2.1 *Faster*

First among the items that contribute to passenger appeal is speed. The whole function of air travel is to provide rapid transportation, and there will always be a market advantage for faster aircraft. To go from the present subsonic regime of most civil aviation to transonic or supersonic flight calls for serious innovation, so let us restrict our reflections here to incremental developments that stay safely within the subsonic regime.

You can see what it will take to go faster if you examine steady flight, where the initial analysis proposed that thrust must be in balance with drag.

$$F_T = F_D \left[= C_D \times \left(\tfrac{1}{2}\rho v^2\right) \times A + \text{ parasitic drag} \right]$$

A fairly sure way to increase speed (v) is to increase thrust (F_T), but that comes at a price. Thrust for most civil aviation comes essentially from burning fuel in jet engines, so unless the engine designers have some radical new schemes for raising efficiency, more thrust requires a heavier power system (probably two or more engines) consuming more fuel.

Exercise 4.9

Suggest two reasons why the use of at least two engines is commonplace in civil aviation.

For a given power system, the thrust it provides can lead to increased speeds (v) only if the combination of factors $C_D \times \rho \times A$ is somehow decreased to compensate. Two factors, the drag coefficient and the wing area, are in the province of wing designers. To reduce the third factor the aeroplane must climb higher and so fly in air of lower density – and before doing that the engine designers must be consulted to be sure that thrust will still be available. The parasitic drag on the fuselage and engine structure can also be expected to increase with speed squared and to scale with air density, cross-sectional area and another drag coefficient. There's more than enough here to keep engineers busy in incremental developments, but perhaps we should be prepared to consider something more radical like integrating the engines, wings and the body into a single lifting surface.

SAQ 4.6 (Outcome 4.4)

If the drag of a particular aeroplane is dominated by the aerodynamic drag of the wings, and the wing is shortened by 2%, it will be able to fly about 1% faster with the same engines. Explain what will happen to the lift that the wings then produce.

4.2.2 Bigger

One of the options for increasing passenger throughput where landing slots are limited is to use aircraft with greater carrying capacity. It is worth asking about the fuselage as a structure to see what constrains its design. Figure 4.31 on pages 116 and 117 shows how an aircraft designer might examine this issue.

The balance between lift and weight suggests a simple design rule for getting a new design off the ground:

$$F_W = F_L \quad \left[= C_L \times \left(\tfrac{1}{2}\rho v^2\right) \times A\right]$$

Is it feasible just to increase wing area in proportion to increased weight? The simple answer is 'yes'. But there are consequences for the whole design. First, since the drag will also increase we will need more thrust to achieve the same speed – more or bigger engines. Then the wing designers will want to be consulted as the weight of the wing will also increase with its area which, together with the extra engine weight, means someone needs to analyse the bending and torsion of the wing structure (Figure 4.32 on pages 118–122). And, while we are reappraising the structures, it is worth enquiring into the ▼**Aerospace materials**▲ (page 123) that must bear the bigger loads. Any improvements in design that save on weight or squeeze a little more lift (C_L) and/or a little less drag (C_D) will be welcome – that has already paid dividends once for Airbus.

Maintaining control of bigger, heavier aircraft will require a re-examination of the stability issues considered in Section 2.8.

SAQ 4.7 (Outcome 4.4)

Seats could be arranged more efficiently in a rectangular cross-section – are there any purely aerodynamic reasons why the fuselage of most passenger jets has a near circular cross-section? Give aerodynamic reasons (not structural).

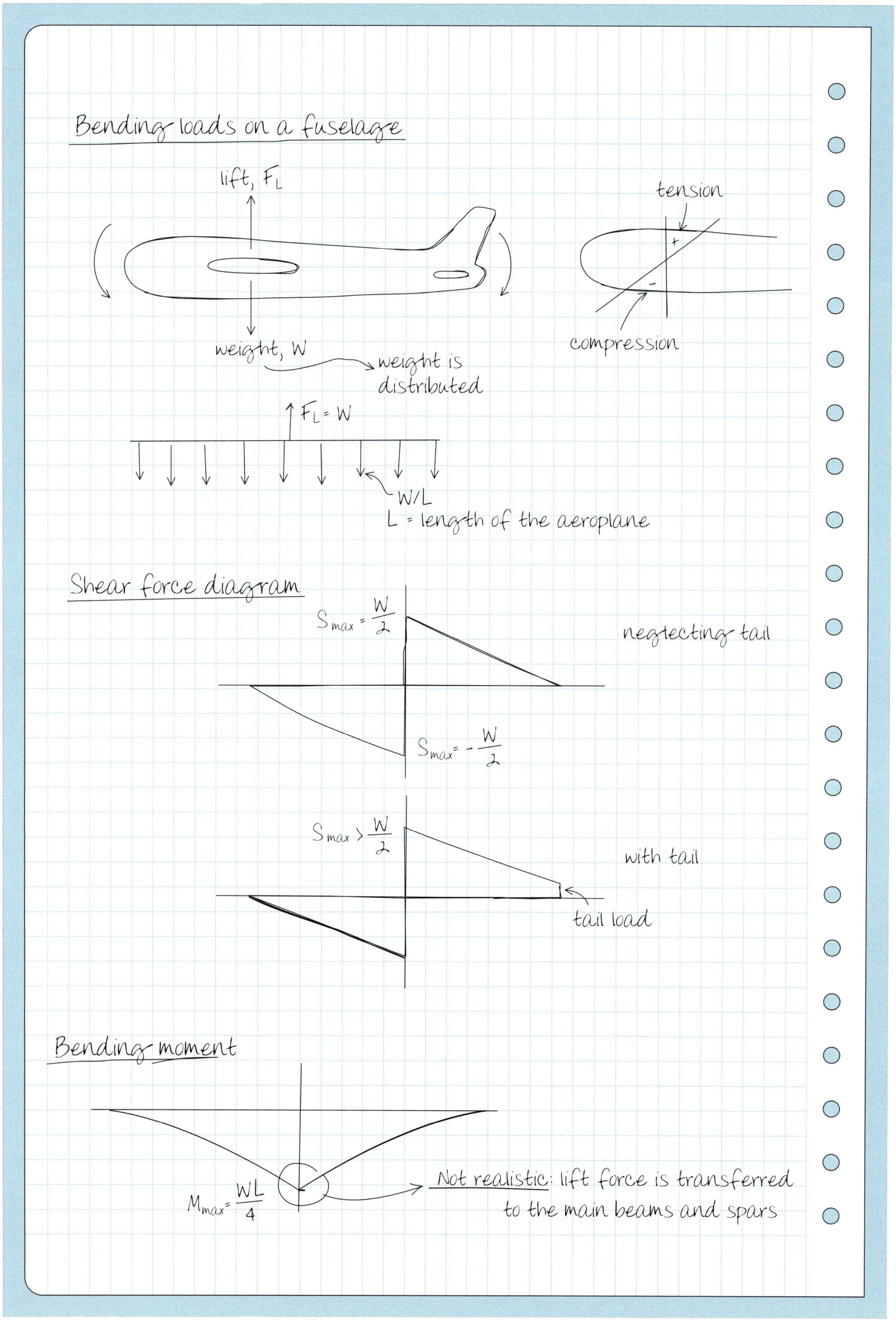

Figure 4.31 The fuselage as a structure – pages from an aircraft designer's engineering notebook

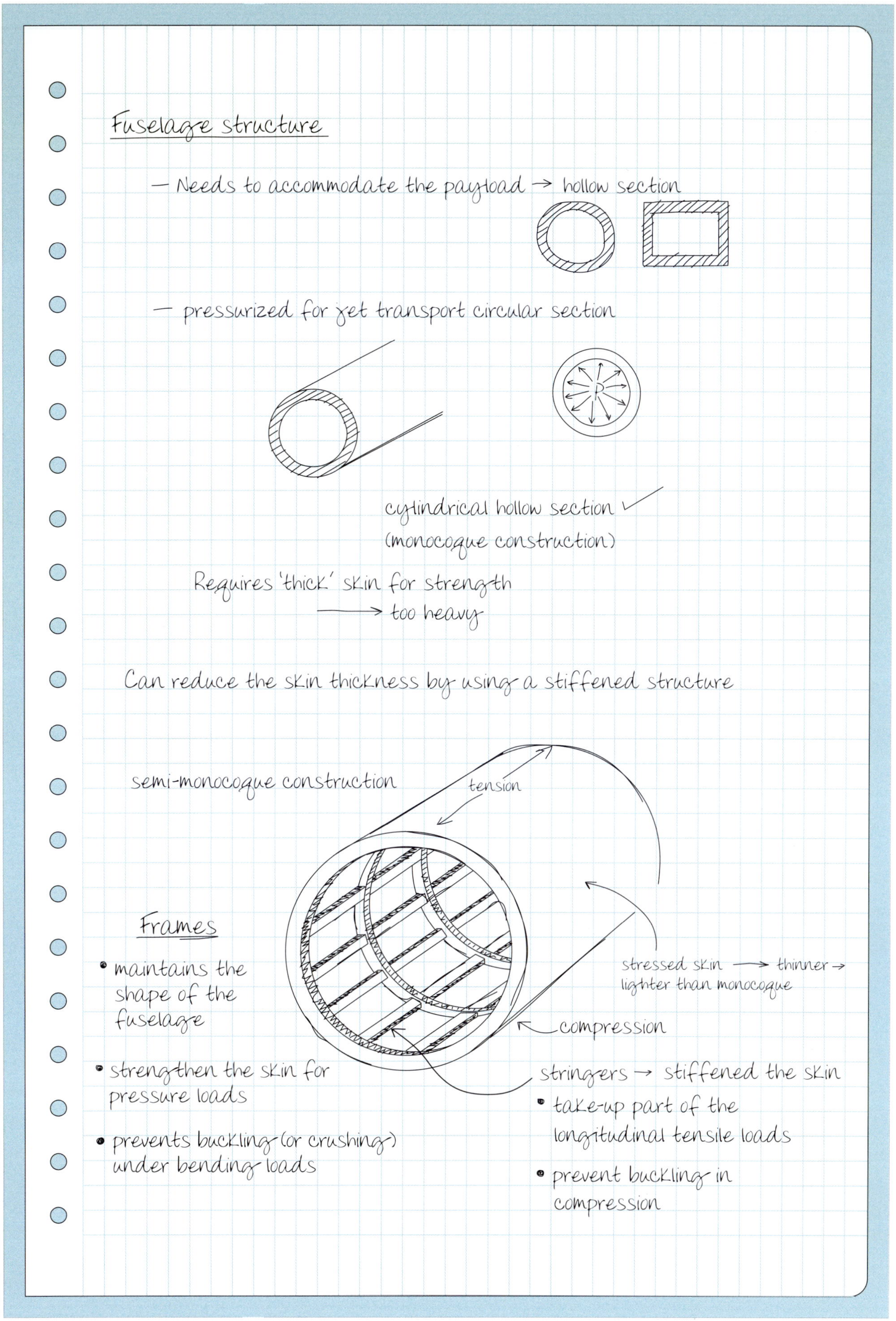

Figure 4.31 continued

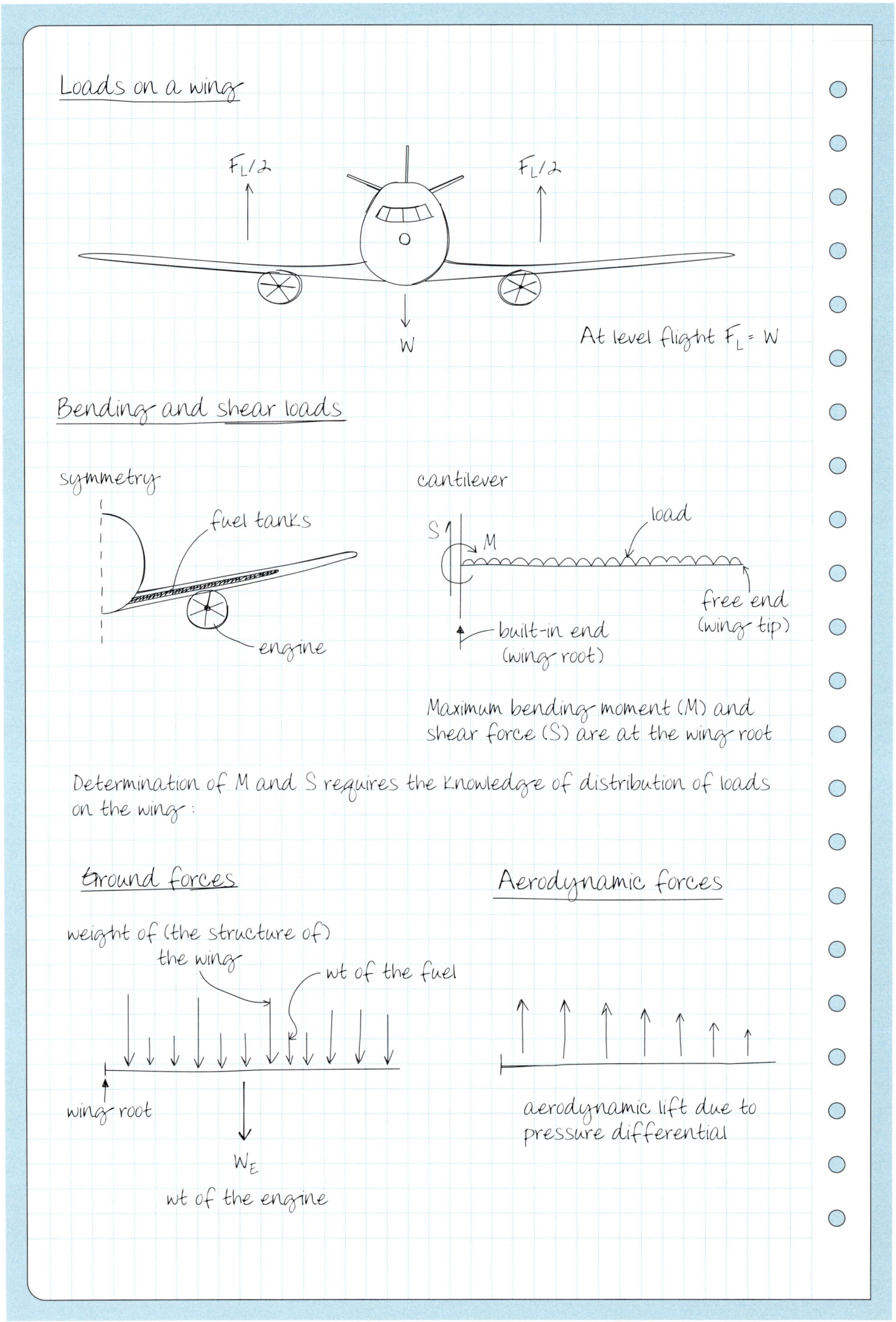

Figure 4.32 Bending and torsion of the wing structure – pages from an aircraft designer's engineering notebook

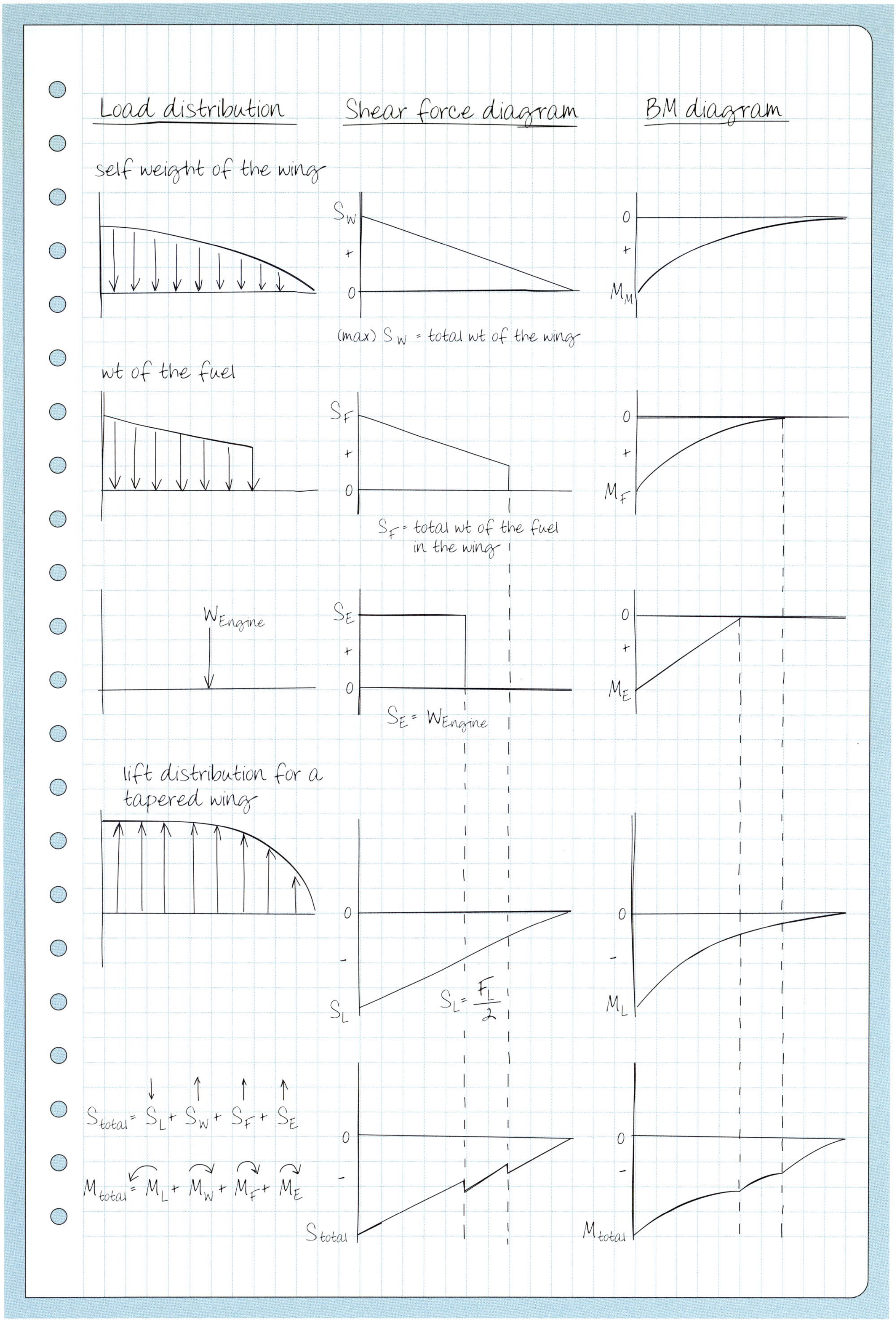

Figure 4.32 continued

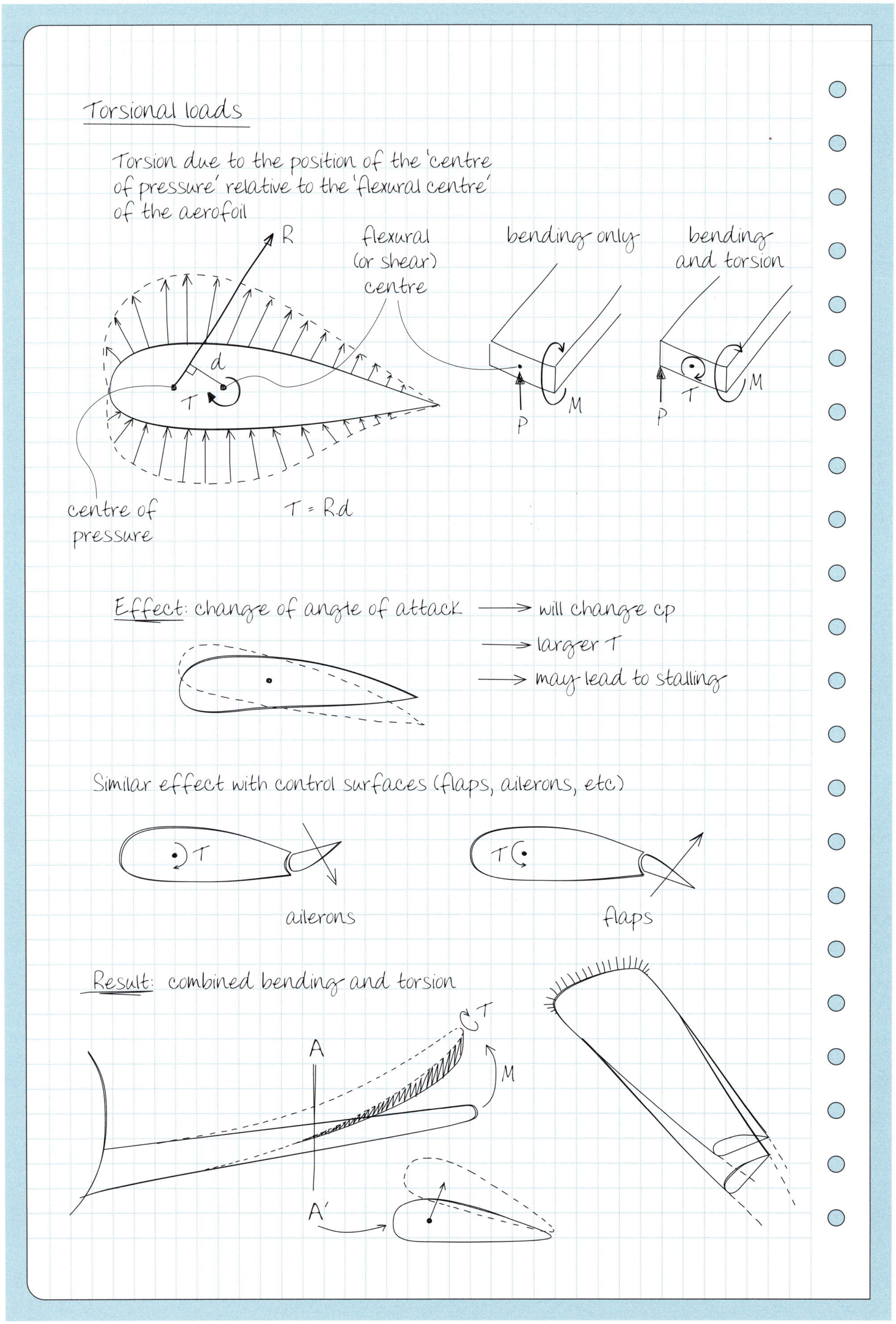

Figure 4.32 continued

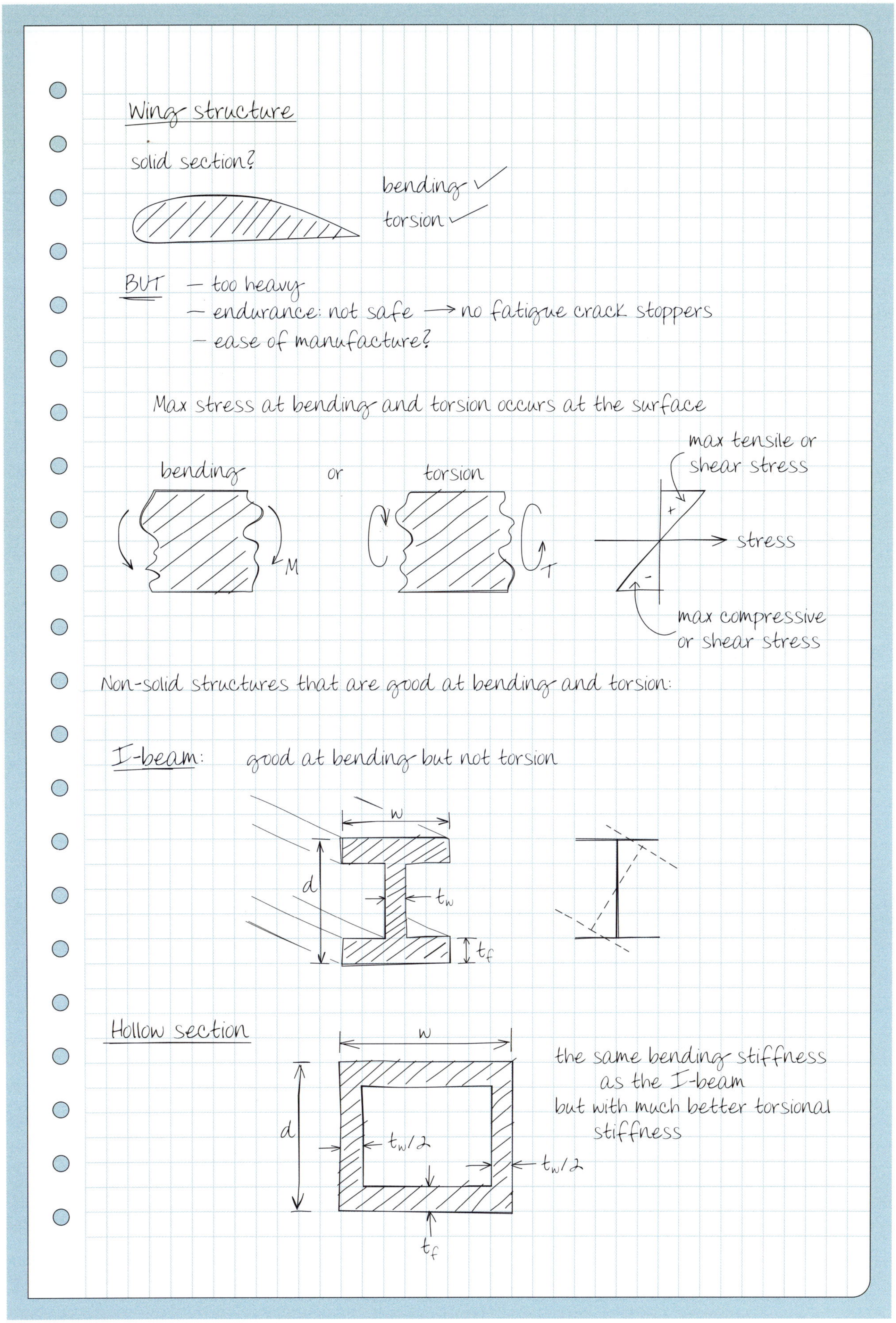

Figure 4.32 continued

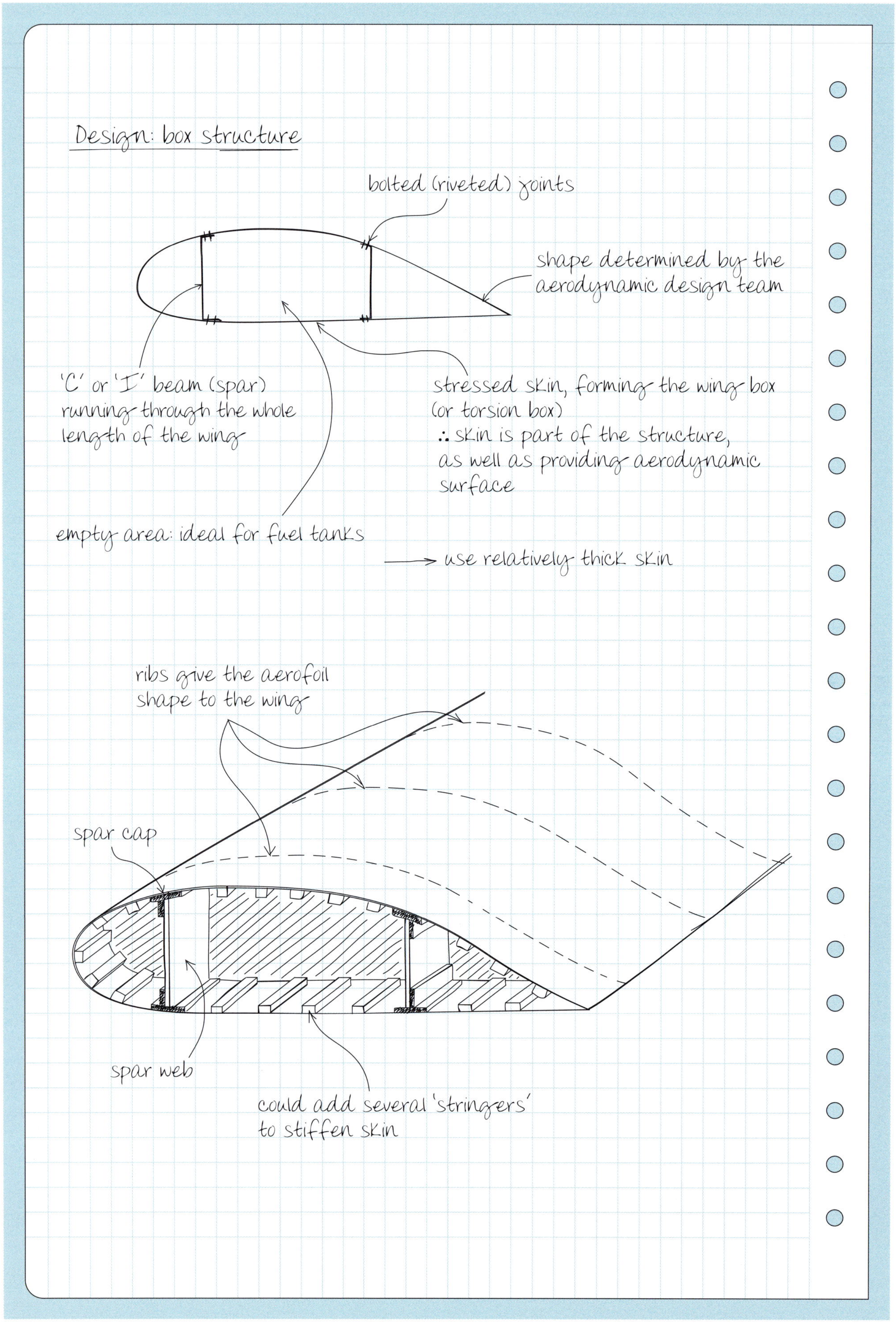

Figure 4.32 continued

▼Aerospace materials▲

The materials chosen to construct the aeroplane must satisfy certain performance requirements at a minimum weight and cost. Therefore, the structural engineer needs to define *merit indices* to compare the suitability of the available materials for a given task. For example, as you saw in Block 1, if the structural engineer was looking for a material for a tie to carry a given tensile load at the lowest possible weight, then tensile strength-to-weight (or density) ratio would be the merit index for that job. The higher the σ/ρ ratio is, the lighter the tie will be. If, in addition, the extension (strain) of the tie under load needed to be small, then elastic-modulus-to-density ratio (E/ρ) would identify the suitable materials. You probably remember from Block 2 how to deduce a merit index for a simple situation such as for a rectangular beam under bending loads.

Aeroplane structures are more complicated than rectangular beams and they are subjected to all sorts of loads; it is fortunate that the designers of aeroplanes go to considerable lengths to design and test them. Nevertheless, it is not surprising that aeroplanes are built from materials for which σ/ρ, E/ρ, $E^{1/2}/\rho$ or $E^{1/3}/\rho$ turn out to be near the highest we can get.

Have a look at Table 4.1 and decide which materials come top for an aeroplane structure.

Interestingly, the aluminium alloys – the most common material in modern aeroplane structures – are often beaten by composite materials (carbon or glass fibre reinforced plastics) and wood (a natural composite). Why, you may ask, aren't such alternatives used more?

Strength and stiffness at low weight are the primary requirements for a material in an airframe structure, but they are not the only factors that the structural engineer needs to consider when selecting materials. Over the page I list some equally important issues affecting the selection process (the list is not exhaustive).

Table 4.1

Material	Tensile strength σ /MPa	Elastic modulus E/GPa	Density ρ /Mg m^{-3}	σ/ρ	E/ρ	$E^{1/2}/\rho$	$E^{1/3}/\rho$
Steel	1150	210	7.85	146	27	1.8	0.76
Aluminium alloy	420	70	2.80	150	25	3.0	1.47
Titanium alloy	900	115	4.44	203	26	2.4	1.10
Wood	35	15	0.70	50	21	5.5	3.52
CFRP (carbon fibre reinforced plastic)	920	150	1.40	657	107	8.7	3.80
GFRP (glass fibre reinforced plastic)	100	45	2.20	45	20	3.0	1.62

- endurance, or the ability of the material to maintain its properties during the service life of the component. These include resistance to degradation or corrosion and fatigue (you will study these in Block 5). The susceptibility of wood to rotting rules out its use in large aeroplanes;
- cost and availability. The cheapest materials in our list are wood and steel. CFRP costs considerably more than aluminium alloys;
- anisotropy and consistency of material properties;
- manufacturing;
- repairability.

4.2.3 Cheaper

Operating costs can be split into two parts. There are flight costs that deal with pilots, **▼Fuel▲**, maintenance, depreciation and insurance. Then there are the less direct costs of the loading equipment, the ramp space, terminal space, cabin attendants, food, management, advertising, selling tickets, etc. Economists can determine the 'per passenger' and 'per flight' costs in a fairly logical way. What might be *charged* 'per passenger mile' is harder to decide on – it could *cost* almost as much to board a passenger (advertise and sell ticket, check-in and load baggage) for a short hop as it does for a long-haul flight. These are the areas of 'low budget' competitions, where engineers rightly fear to tread.

▼Fuel▲

'This is your captain speaking. We have now reached our first cruising altitude of 28 000 ft on our flight from Brisbane to Kuala Lumpur. As our fuel load decreases we shall climb to 35 000 ft as we enter Indonesian airspace. Later we move up to 39 000 ft and eventually to 41 000 ft for the final part of the sector. Our cruising speed will be 820 kph and we shall arrive in KL at 19.30 hours, local time – enjoy your flight'.

I heard this unusually detailed description of a jumbo jet's flight plan a few years ago – why keep going higher, I wondered. The lift and drag equations suggest one answer to my question.

Combining Equations (4.1) with (4.3) and (4.2) with (4.4):

$$\text{lift} = {}^1/_2 C_L A_L \rho v^2 = \text{weight}$$

$$\text{drag} = {}^1/_2 C_D A_D \rho v^2 = \text{thrust}$$

The flight plan has fixed the speed. The weight of the aeroplane gets less as fuel is used, so less lift is required. With v fixed, that means that the pilot can take us higher into lower-density air and still support the weight.

In the lower-density air, the drag at the chosen speed is reduced so less thrust is needed. Throttle back the engines and save fuel!

Air traffic control fix the heights so the altitude goes up in steps rather than a continuous ramp. Presumably the controllers respect the operational need for fuel economy and allow the aeroplane to fly higher and higher as it becomes able to do so.

Concorde was exempt from this restriction. Because Concorde's initial cruise altitude was well over the subsonic jets' vertical-separation-controlled airspace, it could thus climb freely from there until it reached its ceiling. Additionally, Concorde's airways were the only fixed ones over the ocean. And with only three or four flights a day in each direction, the collision probabilities were tiny.

4.2.4 Other factors

An overriding requirement in all aeroplanes is safety. I purposely did not list safety first because it is so self-evident. If someone proposes an innovation that increases speed or payload or otherwise reduces operational costs but compromises safety, it cannot even be considered. There are rigorous international safety requirements that must be satisfied. The requirements cover all safety-related phases of flight including strength, fatigue, stability and control, emergency performance, and emergency design such as fire resistance and control, and evacuation.

Another important factor is comfort. Comfort is affected by a great many things, such as seat width, leg room, seating arrangement, interior noise, vibration, entertainment systems, etc. And, as we established at the start of this part of the block, let's not forget that it is virtually impossible to do any form of engineering that has no consequences, whether direct or indirect, on other people or the environment.

5 POSTSCRIPT

The future of civil aviation relies on the origination of new aircraft and the adaptation of existing models and systems. The development will be centred around the assembler of the airframe – the part to which other bits like wings, engines and tail are ultimately connected. The stakes are high and the competition is fierce.

We set out on this concluding part of Block 3 to explore some of the problems faced by the commercial aviation industry in meeting the steadily increasing demand for passenger miles. That industry needs innovative solutions. It has been necessary to analyse some of the technical aspects of flight, but I hope that you are now better able to appreciate the challenges and the opportunities that exist for civil aviation in the future.

6 LEARNING OUTCOMES

After studying Part 4 of Block 3 you should be able to do the following.

4.1 Recognize the role of engineering in support and infrastructure for civil aviation. (SAQ 4.5)

4.2 Identify commercial factors that have a primary effect on aircraft design. (SAQ 4.5)

4.3 Define lift and drag, and quantities on which each depends. (SAQ 4.1)

4.4 Explain the principles of flight in terms of how aerofoils and engines interact with lift and drag. (SAQs 4.2 and 4.6)

4.5 Explain the stability and control of an aircraft in terms of balanced or unbalanced forces and moments. (SAQs 4.1, 4.2, 4.3 and 4.4)

4.6 Identify certain key engineering advances in the commercial context of civil aviation which in particular account for the following:

(i) Aluminium alloys are used for most airframe materials.

(ii) Most passenger airliners are built around jet engines.

(iii) Alloys used for jet engine turbines combine high temperature strength with low density. (SAQ 4.5)

4.7 Distinguish between the key structural elements of a passenger jet: wings, fuselage, tailplane, skin, spar.

ANSWERS TO EXERCISES

Exercise 4.1

Cost	Engineering materials and engineering dynamics
Aircraft	aerodynamics weight savings by material and structural improvements higher-efficiency engines
Operational (fuel, servicing, pilots and flight crew, ground operations)	maintenance intervals flight control systems ease of access versus aerodynamics aerodynamics resulting in lower fuel consumption
Marketing	—

Exercise 4.2

The aeroplane is:

(a) climbing;

(b) accelerating (forwards);

(c) losing height;

(d) decelerating (slowing down).

Exercise 4.3

The outlet mass flow rate $\dot{m}_o$ must equal that in the constriction $\dot{m}_c$. So remembering that the fluid is incompressible:

$$\rho A_o v_o = \rho A_c v_c$$

Since $A_c < A_o$, it follows that $v_c > v_o$.

Exercise 4.4

To land you must slow down and descend. Just slowing down reduces lift (the v^2 term in the lift equation). To maintain control over the lessening lift, wing area (A) or angle of attack (which affects C_L) can be adjusted. Using the largest wing area available gives the most room for manoeuvre.

Exercise 4.5

(a) The burning of fuel heats the air, increasing its kinetic energy.

(b) The mass flow rate of the fuel must be included in the exhaust.

(c)
$$F = (\dot{m}_{air} + \dot{m}_{fuel}) v_o - \dot{m}_{air} v_i$$

Exercise 4.6

A train is restricted to move on a rail track – so it has just **one** degree of freedom.

A car has relatively more complicated movement; as well as moving forwards (and backwards), it can turn left or right. Although it cannot move sideways directly, the ability of steering provides the means to translate the car in lateral direction (for example, changing lanes on a motorway when travelling at 70 mph). Therefore, its movement is confined to a plane (the surface that it is on – mainly roads); so it has **two** degrees of freedom.

A bicycle, like a car, moves on a plane. But, the cyclist has to keep pedalling and maintain the balance, otherwise the bicycle may fall sideways (rotation about its longitudinal axis). Therefore, it can be said that a bicycle has **three** degrees of freedom.

Aeroplanes travel in three-dimensional space (Figure 4.33); they can *rise* and *sink* along a **vertical axis**, go *forwards* along a **longitudinal axis,** and *sideslip* or *skid* along a **lateral axis**. These axes are orthogonal (perpendicular to each other) and all pass through the aeroplane's centre of mass. As well as these translational motions, aeroplanes can also rotate about these axes; rotation about the longitudinal axis is called *roll*, about the lateral axis is called *pitch* and about the vertical axis is called *yaw* (much like nautical terms). So an aeroplane has **six** degrees of freedom of movement – three rotational and three translational (Figure 4.33).

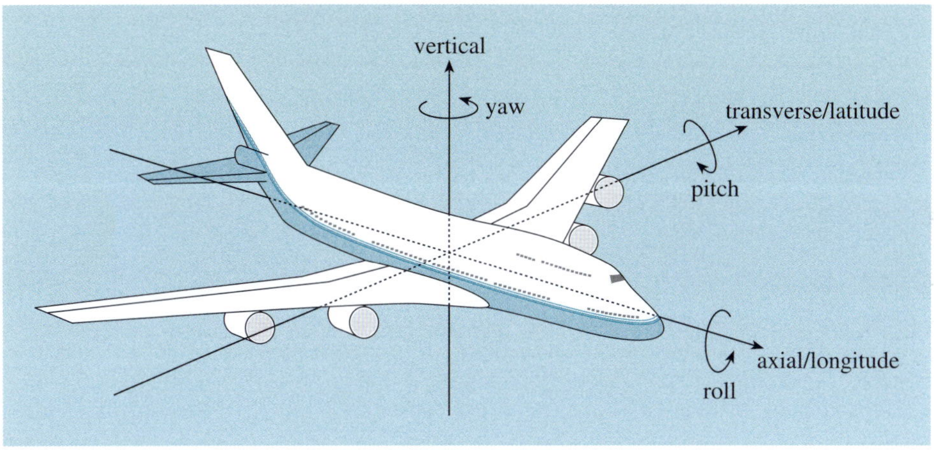

Figure 4.33 Degrees of freedom for an aeroplane

Exercise 4.7

Since engines hang below the centre of mass, their thrust tends to provide a nose-up, turning moment. If the engines were to fail, the nose therefore naturally drops – this enables an unpowered glide to ground. The alternative installation with high-mounted engines could fail catastrophically with nose-up and stall.

Exercise 4.8

You must also recall that whenever an object is accelerating, there should be a force that is responsible for it. Since the acceleration and the force are in the same direction, the force must also be towards the centre. We can find the magnitude of the force using Newton's Second Law:

$$F_C = ma = \frac{mv^2}{r}$$

where m is the mass of the aeroplane. If we write this expression in terms of the weight of the aeroplane ($m = W/g$):

$$F_C = \frac{Wv^2}{gr}$$

This force, called the *centripetal force*, pulls the aeroplane towards the centre of the circle, thus keeping it on its circular path.

Exercise 4.9

Engine failure cannot be completely designed out, so one factor is safety. Either a civil aeroplane must be designed to be a high-performance glider in an emergency or else it is fitted with at least two engines and designed to function sufficiently well on just one.

A second reason concerns balance. It is convenient to hang the engines outside the 'payload' space – one (or two) on each side maintains a static balance. A third engine has been fitted centrally in the tail of some designs of passenger aircraft.

ANSWERS TO SELF-ASSESSMENT QUESTIONS

SAQ 4.1

Since the direction of lift is defined as normal to the airflow (= flight) direction, and the weight is towards the ground,

$$\begin{aligned} F_L &= W\cos 10° \\ &= 200 \times 0.985 \text{ tonnes force} \\ &= 197 \text{ tonnes force} \end{aligned}$$

So the lift force is 197 tonnes force, or 1.93×10^6 N.

SAQ 4.2

This is due to the rather curious definition of 'lift': the component of the upward force, normal to the direction of flight. If we resolve the forces on an aeroplane during climb, with respect to flight direction and equilibrate them:

$$F_L = W \cos \theta$$

$$F_T = F_D + W \sin \theta$$

where θ is the climb angle and F_T is the forward force (thrust) provided by the engines. The second equation tells us that part of the weight is borne by the engines that provide the thrust. Now imagine that the aeroplane is climbing vertically, i.e. $\theta = 90°$. Then, the first equation gives $F_L = 0$ and the second equation gives $F_T = F_D + W$. This means all the lift is provided by the thrust, which is the case for rockets, helicopters and special aircraft that can take off and land vertically, such as Harrier 'jump' jets.

SAQ 4.3

The tailplane influences stability through turning moments. Small forces at large distances are effective, so a large tailplane area is not necessary.

SAQ 4.4

Banking angle

The correct banking angle comes from rearranging Equation (4.5):

$$\tan\theta = \frac{v^2}{gr} = \frac{(400\,000 \text{ m}/3600 \text{ s})^2}{10 \text{ m s}^{-2} \times 1000 \text{ m}} = 1.2346$$

$$\theta = \tan^{-1}(1.2345)$$

So the correct banking angle is 51°.

Required lift

The weight of the aeroplane is:

$$W = mg = 200\,000 \text{ kg} \times 10 \text{ m s}^{-2} = 2\,000\,000 \text{ N}$$

$$F_L = W/\cos\theta = 2\,000\,000 \text{ N}/\cos 51° = 3\,178\,000 \text{ N}$$

So, the wings must produce 3 178 000/2 000 000 = 1.589, or nearly 59% more lift to execute the turn.

SAQ 4.5

Question 1

(a) $10 000 000 000 ($10 billion)

(b) Apart from 'blips' (wars and disasters), the growth in air traffic has been exponential at a rate of 5% per annum for decades. There is total faith that growth will continue, to the extent that the global airline fleet will double from 15 000 aircraft to 30 000 in twenty years. The plane builders' perceived market is 15 000 planes plus replacements over that time.

(c) Twenty years

Question 2

John Lee of Airbus believes that passengers want greater comfort during flying but without greater cost. A 'super-jumbo' can give the space allowing comfort features at similar or lower seat-mile costs.

For Boeing, Randy Baysler explains future passenger expectations, based on experience since deregulation in the USA, as they desire to fly from local airports to wherever they want, whenever and fast.

A380 economics are aimed at high traffic density routes (50 cities worldwide). Boeing's Sonic Cruiser becomes economic because high speed shortens trip times, enabling more sectors to be flown (perhaps by fewer aircraft).

Question 3

It is a matter of history: Boeing has won through (over nearly 100 years!) as the dominant aircraft builder in the USA. Airbus is a consortium of European companies which alone would find it very difficult to compete.

Boeing's plant at Seattle brings all components to the whole aircraft as it is assembled. Mechanics work on all parts of the aircraft simultaneously on a continuously moving assembly line. Improved efficiencies will come from streamlined logistics on the line.

Airbus's facility at Toulouse is a 'final assembly' plant (4% of total job). Pre-assembly of very large sections is dispersed across Europe and they go to Toulouse by air. It would not be economic to bring together the centres of excellence (personnel and facilities) onto one site.

Question 4

(a) Yes, the design is complete and production facilities are being established. A380 is scheduled for service from 2006 – the world's airlines have placed 'a raft of orders'.

(b) 1.49 times the space for 'only' one-third more passengers. Average extra space = $1.49/1.33 \times 100 = 12\frac{1}{2}\%$

Of course, it depends on how the space is deployed, but an extra one-eighth of floor space doesn't change economy flights into QE2 cruising for me.

(c) London and Hong Kong are 'points' and lots of people want to fly between them. It is like this for 50 major cities worldwide, so there are good expectations of high traffic density requiring very large aircraft throughout an extensive long-haul network. Long range at low seat-mile cost means 'big', as the 747 jumbo jet showed.

(In fact, at service inception only thirteen airports are expecting to be able to handle the A380. Runway loadings and terminal facilities need to be upgraded. More airports should become accessible within a few years.)

Question 5

(a) The structure of the aircraft is decided (delta wing, near sonic speed, etc.), but the configurations are still under discussion with potential customers.

Boeing has built a range of plane types from short haul (737) to long haul (747) and Baysler emphasizes the need for aircraft of various sizes to serve the 'point to point' market concept.

Sonic Cruisers may turn out to be a family of aircraft types by the time they reach service in 2008.

(b) 'Faster' makes passengers happy! Airlines can use aircraft on more flights per week if the trip times are shorter. So their schedule may be satisfied by fewer aircraft. That reduces capital expenditure, crew costs and maintenance demands.

(c) Deregulation has permitted civil aviation from/to many more places and with much increased timetable options. Boeing will try to enable the same patterns of passenger behaviour worldwide.

(d) There does seem to be a logical limit – Timbuktu to Milton Keynes will never generate enough clientele to require a scheduled link. But if Airbus sees a 50-city network, Boeing sees some multiple (not specified) of this.

(e) Flying higher means thinner air (less dense). Since lift and drag depend on speed and density, extra speed will give necessary lift in thinner air but will not carry a draft penalty, i.e. will not require more fuel. (Present airliners go higher the further they go because they burn off fuel and get lighter so can get required lift at cruise speed in thinner air with less drag.)

Question 6

Daniel Solon points out that airliners already use different craft for different jobs so both concepts can be winners, particularly if Boeing and Airbus go for different sectors of the market instead of competing.

(Interestingly, in the 1960s Europe went for a fast aircraft. Concorde became a white elephant and the world flew in big Boeings. Has Boeing got it right again?)

SAQ 4.6

The lift and drag associated with a wing scale in exactly the same way with air density, area and speed:

$$F_L = C_L \times \left(\tfrac{1}{2}\rho v^2\right) \times A$$

$$F_D = C_D \times \left(\tfrac{1}{2}\rho v^2\right) \times A$$

So if the reduced wing area and increased speed result in the same drag, they will also lead to the same lift as the original wing at the original speed. The lift and drag coefficients are unchanged by the slight change in design since these are chiefly determined by the aerofoil section.

SAQ 4.7

We have identified no major aerodynamic reasons for circular cross-sections (the arguments are structural). From an aerodynamic perspective, the fuselage holds the wings together and shelters the passengers, at the expense of drag. The fuselage is not an aerofoil.

ACKNOWLEDGEMENTS

Grateful acknowledgement is made to the following sources for permission to reproduce material within Part 4 of this block.

Figures

Figure 4.1: Courtesy of The Shuttleworth Collection; *Figure 4.2:* based on data from Civil Aviation Authority, UK Airports, Annual Statements of Movements, Passengers and Cargo, 2003; *Figure 4.6:* Courtesy of the Lockheed Martin Corporation, USA; *Figure 4.14:* Courtesy of Rolls-Royce Plc; *Figure 4.29:* © Adrian Meredith Photography.

Course team acknowledgements

Part 4 of this block was prepared for the course team by Salih Gungor and Nicholas Braithwaite.